PRAISE FOR **WRITTEN IN STONE**

"Brian Switek's . . . pithy accounts explain how the fossils of everything from Archeopteryx to Tyrannosaurus to Zinjanthropus came to be discovered and interpreted [and his] stories of 19th-century fossil-finders often shed light on current controversies. . . . [An] excellent book."
—MICHAEL SHERMER, *Wall Street Journal*

"In this thoroughly entertaining science history, Switek combines a deep knowledge of the fossil record with a Holmesian compulsion to investigate the myriad ways evolutionary discoveries have been made. It's poetry, serendipity, and smart entertainment because Switek has found the sweet spot between academic treatise and pop culture, a literary locale that is a godsend to armchair explorers everywhere."
—COLLEEN MONDOR, *Booklist*

"Highly recommended."   —*Library Journal*

"Rocks are full of stories. They contain the petrified remains of long-dead animals and in every fossilized bone, scale and track, there are awe-inspiring accounts of the history of life on this planet. Of course, fossils themselves are poor narrators. To uncover their tales, you need a storyteller with an expert's knowledge and a writer's flair. Brian Switek is that storyteller. This is science narrated with maturity, reverence and grace; the epilogue alone is worth the asking price. *Written in Stone*, quite simply, rocks."
—ED YONG, *Discover Magazine*

"Notions of evolution have, for lack of a better word, evolved, and with wonderfully broad strokes science writer and long-time paleontology blogger Switek takes readers on a fascinating historical, scientific and cultural tour of the theory's various incarnations."
—SID PERKINS, *Science News*

"Switek's engaging account may tempt the uncommitted to appreciate how interesting is the underground world, and how the vast storehouses of Earth's strata further our understanding of how life developed. . . . *Written in Stone* is a fine guide to the four-dimensional tapestry of life—the bony bits of it, at least."   —JAN ZALASIEWICZ, *Nature*

"This book will change the minds of those who believe quality science writing is vanishing. Switek has produced prose and paleontological inspiration comparable to the work of the late Stephen Jay Gould. . . . Highly recommended."   —M. A. WILSON, *Choice*

BRIAN SWITEK

WRITTEN IN STONE

THE HIDDEN SECRETS
OF FOSSILS AND THE STORY
OF LIFE ON EARTH

ICON BOOKS

Published in the UK in 2011 by
Icon Books Ltd, Omnibus Business Centre,
39–41 North Road, London N7 9DP
email: info@iconbooks.co.uk
www.iconbooks.co.uk

First published in the USA in 2010 by Bellevue Literary Press,
NYU School of Medicine, 550 First Avenue, OBV 640,
New York, NY 10016

Sold in the UK, Europe, South Africa and Asia
by Faber & Faber Ltd, Bloomsbury House,
74–77 Great Russell Street,
London WC1B 3DA or their agents

Distributed in the UK, Europe, South Africa and Asia
by TBS Ltd, TBS Distribution Centre, Colchester Road,
Frating Green, Colchester CO7 7DW

Published in Australia in 2011
by Allen & Unwin Pty Ltd,
PO Box 8500, 83 Alexander Street,
Crows Nest, NSW 2065

ISBN: 978-184831-262-3

Book design and type formatting by Bernard Schleifer

Printed and bound by
CPI Mackays, Chatham ME5 8TD

For Tracey

Contents

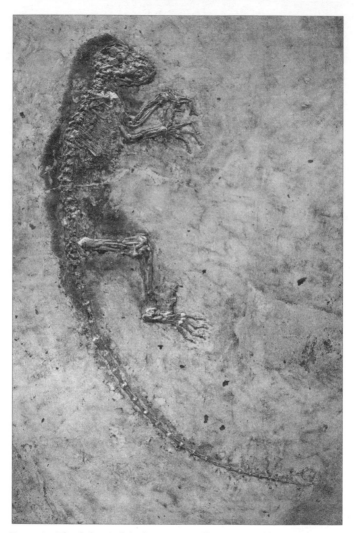

FIGURE 1 – The skeleton of the forty-seven-million-year-old fossil primate *Darwinius masillae*.

Introduction: Missing Links

About thirty years ago there was much talk that geologists ought only to observe and not theorise; and I well remember some one saying that at this rate a man might as well go into a gravel-pit and count the pebbles and describe the colours. How odd it is that anyone should not see that all observation must be for or against some view if it is to be of any service!

—CHARLES DARWIN in a letter to Henry Fawcett, 1861

Let us not be too sure that in putting together the bones of extinct species . . . we are not out of collected fossil remains creating to ourselves a monster.

—SAMUEL BEST, *After Thoughts on Reading Dr. Buckland's Bridgewater Treatise*, 1837

Embedded in a slab of forty-seven-million-year-old rock chipped from a defunct shale quarry in Messel, Germany, the chocolate-colored skeleton lay curled up on its side as if its owner had peacefully passed away in its sleep. Even the outline of the creature's body could be seen, set off in dark splashes against the soft tan of the surrounding stone, but the hands were what immediately drew my attention. Stretched out in front of the body, as if the skeleton was clutching at its slate tomb, each hand bore four fingers and an opposable thumb, all of which were tipped in compressed nubs of bone that would have supported flat nails in life. These were the hands of a primate, one of my close extinct relatives, but was it one of my ancestors?

I had been waiting for days to get a good look at the fossil. My curiosity was initially piqued on May 10, 2009, when the British newspaper the *Daily Mail* announced that the venerable natural history documentary host David Attenborough was preparing to unveil the "Missing Link in human evolution." The full details would be presented in a forthcoming BBC program, the article promised, but as a teaser the piece included a caricature of where our new ancestor fit into our family history. Its lemur-like silhouette stooped at the beginning of a short parade of human evolution conducted through our primate antecedents to us.

Further details about the fossil were difficult to dig up. A May 15, 2009, piece by the *Wall Street Journal* provided little new information other than that the discovery would be unveiled the following Tuesday during a New York City press conference coordinated with the release of a descriptive paper in the journal *PLoS One*. This made sense of a nauseatingly overhyped press release I had received the day before which shouted "WORLD RENOWNED SCIENTISTS REVEAL A REVOLUTIONARY SCIENTIFIC FIND THAT WILL CHANGE EVERYTHING." The fossil would be presented with all the pomp and circumstance due a newly discovered and long-lost family member, but I did not care as much about the public ceremonies as the scientific paper. I wanted to know if the evidence supported the fantastic claims being bandied about in the newspapers.

I had hoped that *PLoS One* would send out an embargoed version of the paper so that science writers like me could brace for what was promised to be an earthshaking announcement. This is a standard practice in which a journal distributes papers to science writers a few days early so that stories can be prepared (with the understanding that no one will break the story until the embargo lifts), and *PLoS One* had used it for many of its major publications. No such luck. Science writers would have to wait for the grand unveiling like everyone else.

When the paper was finally released I felt simultaneously overjoyed and underwhelmed. The petrified skeleton—named *Darwinius masillae* by the authors of the study in honor of Charles Darwin—was the most beautifully preserved primate fossil ever discovered. The remains of prehistoric primates are rare to begin with; most of the time paleontologists find only teeth and bone fragments. But *Darwinius* was exquisitely preserved with hair impressions and gut contents in place. Even the famous skeleton of our early relative "Lucy" was far less complete. By any estimation, this first specimen of *Darwinius* was a gorgeous fossil.

Despite the intricate nature of the fossil's preservation, however, the evidence that *Darwinius* was even close to our ancestry was flimsy. The paper confirmed that it was a type of extinct primate called an adapiform, and while they were once thought to be good candidates for early human ancestors more recent research showed that lemurs, lorises, and bush babies are their closest living relatives. In order to change this consensus *Darwinius* would have to exhibit some hitherto unknown characteristic that affiliated it more closely with early anthropoid primates (monkeys and apes, including us), but the authors did not make a good case for such a connection. There was no trait-for-trait comparison of *Darwinius* with other living and fossil primates that would have supported the status of "ancestor" that early reports had given it.

None of this hindered the fossil's bombastic media debut. In public the fossil was called "Ida" after the daughter of one of the paper's authors, paleontologist Jørn Hurum, and Hurum introduced Ida as our unquestionable ancestor. He proclaimed that *Darwinius* was "the first link to all humans . . . the closest thing we can get to a direct ancestor." Some of his co-authors were equally given to hyperbole. Paleontologist Philip Gingerich compared *Darwinius* to the Rosetta Stone, and lead author Jens Franzen stated that the effect of their research would "be like an asteroid hitting the Earth." A pair of high-profile documentaries, a top-notch Web site, a widely read book, and dozens of early media reports drove home the same message; Ida was the "Missing Link" that chained us to our evolutionary history.

New York Times journalist Tim Arango beautifully described this tidal wave of publicity as "science for the Mediacene age." In an instant Ida was everywhere. After seeing the fossil plastered all over the news and even in a customized Google logo I half expected to find promotional "The Link" breakfast cereal at the supermarket. The premiere was just as well orchestrated as that of any Hollywood blockbuster, but unlike most big-budget films there was no buzz leading up to the big event. Outside of the early reports from the *Daily Mail* and *Wall Street Journal* barely a peep was heard about Ida before her debut.

Scientists and journalists who were not content with regurgitating the approved press releases scrambled to dig up the glorified lemur's backstory. Something was not right. The public was being sold extraordinary claims about Ida before anyone had a chance to see if the science held up to scrutiny. It was the scientific equivalent of not screening a film for review by critics but promoting the movie as the greatest since *Casablanca*. Hurum was unapologetic about this media strategy. "Any pop band is doing the same thing," he dodged. "Any athlete is doing the same thing. We have to start thinking the same way in science." But, as Hurum well knew, there was much more to it than that. As reports started to trickle in from independent sources it quickly became apparent that Ida had been groomed for stardom almost from the very start.

When the fossil pit in Messel, Germany, coughed up Ida it was on its way to becoming a garbage dump. The quarry had been a shale mine for years. Numerous exquisitely preserved fossils had been discovered there, but after the mining operations stopped in 1971 the government made preparations to turn it into a landfill. Amateur fossil hunters knew their time was limited. They picked over the site to remove whatever they could, and in 1983 one of the rock hounds split open a slab of shale to discover Ida's skeleton.[1] There were two parts: a mostly complete main slab; and a second slab that, because of the angle of the split, was missing some of the bones of the head, leg, and torso. Rather than stitch them back together, Ida's discoverer hired a fossil preparator to fill in the details of the "lesser half," using the more complete slab as a guide.

Such a discovery was too valuable to just give away to science, and the half-real, half-fabricated slab was sold to the Wyoming Dinosaur Center in 1991. Perhaps the fossil should have been called "*Caveat emptor*" at this point; not only was the purchased slab partially faked, but the parts that were real were not especially helpful in determining what kind of primate it might have been. The specimen sat virtually unnoticed in the Wyoming museum. The other slab stayed in private hands. Scientists had no idea it existed.

By 2006, however, it was time to sell Ida's better half. Her owner (who has remained anonymous) sold it to the German fossil dealer Thomas Perner, who in turn offered it to two German museums, but Perner's asking price was so high that neither institution could afford the fossil. Private collectors have deeper pockets than museums, though, so Perner decided to bring a few high-resolution photos to the Hamburg Fossil and Mineral Fair to show to some of his previous clients, including University of Oslo paleontologist Jørn Hurum.

Upon seeing the fossil, Hurum was instantly enthralled. He had to have it. The trick would be raising the $1,000,000 Perner was asking. He could not afford this on his own but hoped his university could help foot the bill. Eventually they reached a deal. The college would dole out a total of $750,000 in two payments: half the asking price once the fossil was in Hurum's hands and the other half when he were sure of its authenticity. The tests confirmed that, unlike its complement, the slab had not been forged, and by the beginning of 2007 Hurum finally had his fossil "Mona Lisa."

But Hurum was not a primate expert. Most of his scientific work had focused on dinosaurs and extinct marine reptiles. To make up for this lack of expertise he put together what he would later call an international "dream team" of fossil primate specialists; Jens Franzen, Philip Gingerich, Jörg Habersetzer, Wighart von Koenigswald, and B. Holly Smith. Each scientist brought different strengths to the team, but the inclusion of Franzen was especially important. Franzen had described the other half of Ida's skeleton during the 1990s, and once it was realized that the two slabs were halves of the same fossil they were reunited.

Hurum also had bigger things in mind. At the time he acquired Ida,

Hurum was working with the media company Atlantic Productions on a documentary about the remains of a 147-million-year-old, fifty-foot-long carnivorous marine reptile given the B-movie moniker "Predator X." The company had jumped at the chance to document the study of one of the largest marine predators that ever lived, and Hurum approached them about Ida. The company reps were just as taken with the primate fossil as Hurum was. Sea monsters were interesting, but a potential human ancestor was even better. Plans for the two documentaries, the mass market book, and all the other details of the public release began to coalesce.

Team member Philip Gingerich would later lament, "It's not how I like to do science." With the May 19, 2009, debut date set far in advance the scientists had to rush to get their description of *Darwinius* completed in time. This presented a substantial hurdle. To be published in a reputable scientific journal research must go through a process of peer review in which the original paper is sent for comment to academics in the same field. Based upon these independent assessments the journal then decides to either publish or reject the paper, and even if the paper is not rejected it might still require changes prior to final acceptance. The process can drag out for months or even years, and since the first complete version of the *Darwinius* paper was completed in the early months of 2009 the researchers did not have much time left.

As the open-access journal *PLoS One* had earned a reputation for a speedy review process, it seemed like the best choice. The manuscript was submitted in March, but it could not immediately be accepted. According to one of the reviewers, fossil primate expert John Fleagle, the paper made the extraordinary claim that *Darwinius* was a human ancestor without supplying sufficient evidence. This conclusion was toned down, and in the next draft the authors suggested that *Darwinius* might be closely related to the ancestors of anthropoid primates instead. Nevertheless, the plans to herald Ida as the "missing link" to the public remained in place, and despite the heavy involvement of the media companies, the scientists declared no competing interests in the paper.

The paper was finally accepted on May 12, 2009, just one week before it was set to be released. With the contents of the paper finalized, the *PLoS One* employees went into overdrive to get the paper prepared for Ida's debut. They managed to finish their work by May 18, but on behalf of the media companies the authors asked that the paper not be released to anyone until the press conference the next day.[2] The journal acquiesced. Atlantic Productions was given full control over how Ida would be presented.

When this convoluted tale of black market fossil deals, pervasive media control, and overhyped conclusions burst onto the public scene scientists were aghast. There were so many controversial points it was difficult to know where to start, but the most prominent was Ida's being hailed as our great-great-great-great- . . . -grandmother. By all appearances *Darwinius* had been believed to be a human ancestor from almost the start. This was not good science and, in truth, the peer review of Ida had only just begun.

A hypothesis or conclusion announced in a scientific paper is not ironclad law. Publication is just an intermediate step in fostering our understanding of nature, and a hypothesis will stand or fall according to the ensuing debate. The case of *Darwinius* was no exception. It was clear that the team of scientists had not done the essential work to support the claims they were making in public, and within a few months a new study would put Ida in her proper place.

In 2001, five years prior to the sale of *Darwinius* to Hurum, paleontologist Erik Seiffert and his colleagues were searching for fossils in the thirty-seven-million-year-old sediments of the Fayum desert of Egypt. During that part of earth's history the Fayum hosted a lush forest inhabited by a mix of early anthropoids and representatives of other now-extinct primate groups. Among the fossil scraps Seiffert and his peers collected in 2001 were the jaw fragments and teeth of a lemurlike primate. The distinctive shape of a mammal's teeth is so closely tied to its feeding habits that a handful of teeth can be more useful in determining its closeness to another mammal than scattered bits of ribs, limbs, or vertebrae.

The Fayum team spent years piecing together the bits of the primate they had found, but in the wake of the Ida fallout Seiffert and colleagues Jonathan Perry, Elwyn Simons, and Doug Boyer resolved to do what "team *Darwinius*" had not. They compared 360 characteristics across 117 living and extinct primates, including *Darwinius*, through a methodology known as cladistics.

The logic behind the technique is simple. The goal is to create a tree of evolutionary relationships based upon common ancestry, and to do this scientists select the organisms to be scrutinized, choose the traits to be compared, and document the character state of each trait (i.e., whether the trait is present or absent). Once all this information is compiled it is placed into a computer program that sifts through the data to determine which organisms are most closely related to each other on the basis of shared, specialized characteristics inherited from a common ancestor. Anthropoid primates and tarsiers, for example, have a partition of bone which closes off the back of the eye, whereas lemurs and lorises lack this closure. The fact that *Darwinius* lacked this distinctive plate of bone behind its eye, among other characteristics, associated it closer with lemurs and lorises than anthropoid primates.

No single trait overrides all the others, though. Some traits evolve more than once in different lineages or are secondarily lost among some members of a group, so it is better to select numerous traits rather than just a handful. Each evolutionary tree produced is a hypothesis that will be tested against additional evidence, but cladistics has the advantage of forcing scientists to fully present the data they use in the process. Even if the resultant tree is thought to be incorrect, scientists can at least look at the data to pinpoint what might have skewed the results. This kind of self-correction is not possible when ancestors and descendants are lined up simply on the basis of what looks right.

The results of the analysis Seiffert and his team conducted were published in the journal *Nature* on October 21, 2009, just over five months after *Darwinius* was announced. There were a few surprises. Despite living thousands of miles and ten million years apart, the primate from the Fayum, which they named *Afradapis longicristatus*, was a very close relative of *Darwinius*. They were definitely both adapiforms, but they were unusual ones.

Both *Darwinius* and *Afradapis* had traits that had traditionally been thought to be indicative of anthropoid primates, such as the fusion of the lower jawbones where they meet in the middle. This is a key trait seen in living monkeys, not lemurs, and if we had only living primates to compare *Darwinius* to then we might think that adapiforms really were ancestors of anthropoids. The problem is that some of the earliest anthropoid primates known, such as *Biretia* and *Proteopithecus*, do not share these same "anthropoid features." These traits evolved independently among later anthropoids in a case of convergent evolution. For *Darwinius* to be an anthropoid ancestor its descendants would have had to lose some traits, such as the fused lower jawbones, only to have those same traits evolve again later among its descendants. There was no evidence to suppose that such a thing had happened.

This conclusion was supported by the evolutionary tree Seiffert's team produced. Not only did *Darwinius* and *Afradapis* group closely together on the basis of their shared characteristics, but they were about as distantly related to early anthropoids as it was possible to be. Their closest living relatives are the lemurs and lorises, not monkeys. (Though they actually were most closely related to other forms of primate that are now entirely extinct.) As expected, it was the tarsiers and their extinct relatives that were most closely related to anthropoids. Ida had unceremoniously been dethroned.[3]

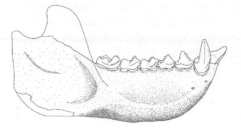

FIGURE 2 – The lower jaw of *Afradapis longicristatus,* reconstructed on the basis of multiple specimens. So far, it is all that is known of this fossil primate.

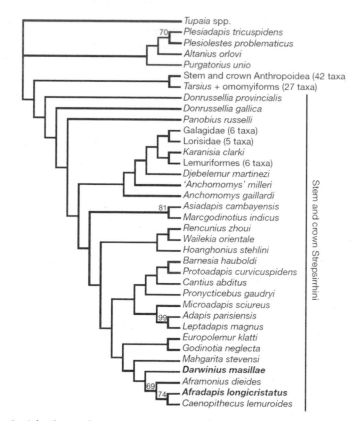

FIGURE 3 – A family tree of primates as produced by the cladistic analysis run by Seiffert and colleagues. Not only does *Darwinius* fall near *Afradapis,* but both are confirmed as extinct relatives of lemurs far removed from anthropoid primates.

Her backers were not pleased. Distancing himself from the headline-making claims of a few months before, Hurum stated that *Darwinius* could still belong to a "stem group" from which early anthropoids evolved. After all, the skeleton of *Darwinius* was much more complete, and according to Hurum it contained some anthropoid characteristics that could not be seen in the incomplete remains of *Afradapis*. Gingerich was similarly unimpressed. He asserted that the anthropoid traits seen in *Darwinius* were not convergences at all; Ida had monkey-like traits because she was closely related to monkeys. Though the *Afradapis* paper presented a much better supported hypothesis for what the primate family tree looks like, it was hardly the last word on the matter, either. Hurum promised that an independent cladistic analysis of *Darwinius* was already being planned.

I watched this back-and-forth from the periphery. As a writer there was not much I could directly contribute to the scientific discourse, but I was hooked by the drama surrounding Ida.[4] I couldn't help but wonder why this petrified primate had caused such a fuss. If Ida had been presented in her proper evolutionary position, as a unique relative of living lemurs, this whole media kerfuffle probably would not have happened. Therein was my answer.

No matter how much we learn about nature there are some questions our species continually grapples with. Why are we here? How did we get to be this way? Where are we going? Maybe these questions sound a bit trite, but if that is so it is only because they are timeless queries that have been difficult to answer. We desperately want to know where we came

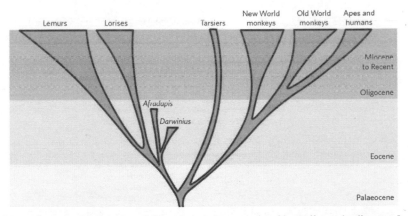

FIGURE 4 – A simplified version of the evolutionary tree produced by Seiffert and colleagues. It shows a deep split among early primates, with *Darwinius* and *Afradapis* being on the side that gave rise to lemurs and lorises, not anthropoids.

from, where we are headed, and, as phrased by novelist Douglas Adams, the "Ultimate Answer to Life, the Universe and Everything."[5]

The answers to these questions have traditionally been supplied by religion. We have been created and sustained thanks to God's will, so the story goes, making us the most privileged thing in all Creation. Even if we feel lost and isolated we can still believe that there is an inherent purpose and direction to life, a beginning and an ending.

But during the past 150 years these existential questions have taken on new inflections. There might not be a universal answer to "Why are we here?" that provides us with a driving sense of purpose, but an understanding of the quirks and contingencies of evolution allows us to meaningfully understand how we came to be as we are. This was made possible by the work of Charles Darwin in the middle of the nineteenth century. He was not the first person to consider evolution, nor was he the only Victorian naturalist to provide evidence for it, but through his 1859 masterwork *On the Origin of Species* Darwin popularized a new view of life in which a past far beyond the oldest remnants of human history could help us understand our place in nature. We are inextricably tied to what has come before.

Our preoccupation with origins made the search for fossil ancestors among the most pressing preoccupations of naturalists. If life had truly been transforming over an incalculable amount of time, then the bones of our distant ancestors, as well as forerunners of every other living species, should speak to us from the earth. This hypothesis was a bit of a gamble for Darwin. Geology and paleontology had been essential to the formation of his evolutionary theory, yet the records of deep time had, prior to 1859, failed to provide the continuous, graded chains of fossils linking the present to the past. While Darwin was correct that the fossil record was an archive "imperfectly kept," full of gaps and discontinuities, ultimately it would have to provide the solid proofs of the theory he had based on observations of living animals.

The rarity of these fossil proofs of evolution vexed naturalists. In an 1868 address on the evolution of birds from reptiles Darwin's ally Thomas Henry Huxley likened the state of affairs to a landowner who, despite his claims, could not produce hard evidence that he really owned the property at all:

> If a landed proprietor is asked to produce the title-deeds of his estate, and is obliged to reply that some of them were destroyed in a fire a century ago, that some were carried off by a dishonest attorney, and that the rest are in a safe somewhere, but that he really cannot lay his hands upon them; he cannot,

I think, feel pleasantly secure, though all his allegations may be correct and his ownership indisputable. But a doctrine is a scientific estate, and the holder must always be able to produce his title-deeds, in a way of direct evidence, or take the penalty of that peculiar discomfort to which I have referred.

Naturalists would have to supply these "title deeds" if the fact of evolution was to be established. The theoretical question of whether evolution was driven by natural selection or some other force would be debated for decades, but the fossil record held the most immediate potential of supplying solid evidence that evolution was real.

This want of ancestors is what allowed the *Darwinius*-for-ancestor lobby to enthrall the public. The fossil record does not contain a complete roll of every living thing that ever lived. It is rare that a living thing dies in circumstances amenable to fossilization, and even among this fossil pool the remains of many organisms are destroyed by geological processes. Of this fraction of a fraction only a very few specimens exist in rocks accessible to scientists, and of that tiny slice fewer still are collected and studied. The discovery of any fossil with transitional features that helps us understand the transformation of one form into another is cause for celebration, and most celebrated of all are those that connect familiar animals to their extinct forerunners.

The fossil forms which bridge the gap between one group of organisms and another have popularly been called "missing links" (and this is especially true of the search for our own ancestors). This is an unfortunate misnomer that reveals the ancient origin of the phrase as well as the biases that run though it. Indeed, the idea of missing links originally did not contain any evolutionary significance at all. During the Middle Ages Christian scholars thought that life was organized according to a hierarchical scale of natural productions ranked from "lower" to "higher." This was the Great Chain of Being, and it was a static arrangement that reflected the virtues of Creation: plentitude, continuity, and gradation.

Since God was benevolent and omnipotent He had created everything that was possible.[6] Ours was the best of all possible worlds, one of magnificent plenitude, but there was an order to the diversity of nature. In the continuous, unbroken hierarchy anything in nature could be linked to another by recognizing their shared characteristics. A rock had existence, while a plant had both existence and life, and the ability of animals to move around on their own placed them above plants. And so the rankings went, all the way from pebbles up to the Almighty, with humans representing the highest point of the "animal Creation." Our kind was a step above other animals but one below angels, beings possessed of a heavenly infused soul but still subject to animal urges.

Despite the certainty that God had ordered creation according to these laws, however, there were breaks in the chain. Among the most troublesome was the one between humans and the vulgar monkeys (which, for many medieval Christians, represented what a life of sin could lead to). Monkeys were clearly similar to humans but far too low to be on the rung right below us. Between us and them there should have been a humanlike being that lacked a soul, but for centuries this missing link remained elusive.

This view of nature was later co-opted into ideas about evolution. By the beginning of the nineteenth century the Great Chain of Being ceased to be a useful concept to organize nature, but vestiges of it still remained. The vertical dimension of the hierarchy, rather than representing only the rank of living organisms, was impressed onto the geological timeline. Fish appeared before amphibians, which preceded reptiles, which in turn gave way to an Age of Mammals capped by the appearance of our own species. The story of evolution still presented a chain of beings connected through a series of intermediate links, and it was among fossil vertebrates that the first of these intermediate forms were found. In his 1870 address as president of London's Geological Society, Huxley stated that "when we turn to the higher *Vertebrata,* the results of recent investigations, however we may sift and criticize them, seem to me to leave a clear balance in favour of the doctrine of the evolution of living forms one from another." Fossil vertebrates provided some of the most compelling evidence for evolutionary change, and it was not surprising that some scientists interpreted the succession of these forms to represent life's progress.

This underlying thread has given rise to some of our most iconic evolutionary images. The March of Progress from early primate to human is one, but the same imagery has been employed for the evolution of horses, elephants, the earliest terrestrial vertebrates, early mammals, birds, and whales. As transitional forms have been found they have been strung up in temporal sequences to show the progressive transformation of the archaic into the modern. This interpretation might not be explicit, and perhaps it is even outright denied by the presenters of these diagrams, but such illustrations leave little doubt that the biases inherent in the Great Chain of Being remain with us even today.

And this drive toward progress implies the question of what might come next, particularly for our own species. What might our descendants be like a thousand, a million, or ten million years from now? If the past presents us with a tale of progress from "primitive" to "advanced," then what might the future hold for us? What is the next evolutionary step? There is no way to tell. It is impossible to predict how our species might be adapted, but the annals of science fiction reveal our expectations. It is no coincidence that in popular culture, from Hollywood films to discussion boards run by UFO conspiracy nuts, technologically superior aliens are envisaged as having large heads stuffed with enormous brains and frail humanoid bodies.[7] They are species that have advanced to the point where body is sublimated to mind, and they act as proxies for what many expect our species to become given enough time. As hypothetical creatures that live more in the mental realm than the physical, they occupy the place once inhabited by angels, above humans but below God, on the Great Chain of Being.

The irony of this view is that Darwin envisioned evolution as producing a wildly branching tree of life with no predetermined path or endpoint. It is significant that the only illustration in *On the Origin of Species* is not a revised version of the Great Chain of Being, but a series of branches embedded within greater branches, all connected by common ancestry. With a sufficiently complete fossil record it is possible to trace the evolution of particular forms according to direct lines of descent, but doing so requires that neighbouring branches containing close relatives be lopped off. And the further back in time we go, the more relatives we have to ignore.

Any paleontologist worth their salt knows this well. Yes, it is possible to line up a series of forms representing what our direct ancestors looked like at different points over the last six million years or so, but to do so would require that we ignore other types of early humans that lived alongside our ancestors such as the heavy-jawed robust australopithecines and our sister species, the Neanderthals. Even before that, our anthropoid ancestors were just one twig of a more diverse evolutionary bush that coexisted with other kinds of primates such as *Afradapis* and tarsiers. To focus solely upon our ancestors is to blind ourselves to our own evolutionary context.

But why consider fossils at all? In the introductory chapters of his 2004 tome *The Ancestor's Tale* Richard Dawkins stressed that "dead men tell no tales." We might be just as well off in our understanding of evolution if not a single fossil even existed:

> In spite of the fascination of fossils, it is surprising how much we would still know about our evolutionary past without them. If every fossil were magicked away, the comparative study of modern organisms, of how their patterns of resemblances, especially of their genetic sequences, are distributed among

species, and of how species are distributed among continents and islands, would still demonstrate, beyond all sane doubt, that our history is evolutionary. Fossils are a bonus. A welcome bonus, to be sure, but not an essential one.

But this dim view of paleontology is not accurate.[8] During the past thirty years scientists have seen the emergence of a new, synthetic paleontology that is giving us an unprecedented look at the machinations of evolution.

Scientists such as Stephen Jay Gould, Niles Eldredge, Steven Stanley, Elisabeth Vrba, David Raup, and Jack Sepkoski led the charge. Starting in the 1970s these paleontologists questioned the popular interpretation of evolution as a slow and steady process in which species were constantly evolving by tiny steps. Their research was not in conflict with evolution by means of natural selection, but the patterns of the fossil record were far more haphazard than had been expected on the basis of genetics. This precipitated a 1980 conference in Chicago where some of these ideas could be hashed out with biologists, such as John Maynard Smith, who favored a smoother evolutionary pattern. The tension was felt by all. After a presentation by embryologist George Oster about how developmental quirks constrain the forms organisms can take, Maynard Smith responded that scientists like himself had already considered the idea and dispensed with it as being of little importance. The paleontologists and other biologists who were questioning what was commonly accepted were only reinventing the wheel. To this Oster retorted, "You may have had the wheel, John, but you didn't ride away on it."

Paleontologists were ready to hop on and see where they could go, and while relations between paleontologists and neontologists (biologists who work with living organisms) were tense to start with, the debates between them started to feed cross-disciplinary collaborations. Slowly, paleontologists began to incorporate discoveries from molecular biology, genetics, and embryology into their work. This allowed paleontologists not only identify to patterns of change, but to begin to understand how such changes in form might have been caused. The discovery of preserved soft tissues from prehistoric creatures from Neanderthals to mammoths to *Tyrannosaurus* have even opened a new field of study centered on the recovery and study of ancient molecular materials. Comparative anatomy and geology still form the core of paleontology, but the science has embraced information and techniques from a variety of disciplines, thus allowing scientists to test their ideas about life's history through the combination of multiple lines of evidence.

The coalescence of this new paleobiological synthesis coincided with the discovery of many new transitional fossils and the reappraisal of many old ones. Fossils that scientists knew had to exist but had been missing, such as land-dwelling whale ancestors and feathered dinosaurs, were found, while well-known lineages, such as horses and elephants, were revealed to show a wildly branching pattern of diversity rather than a straight line of progress. Even among our own ancestors, what had once been supposed to be a single chain of ancient humans was suddenly split and split again by new discoveries, so much so that at the turn of the twenty-first century no less than three fossils were in competition for the designation "earliest human." With the development of the new paleobiology and more complete collections of transitional fossils, paleontologists began to piece together a better understanding for how life changed through time. Paleontology is not just a bonus; it is among the most essential of evolutionary disciplines.

Fossils do not speak for themselves, however, and the history of science fleshes out the context in which new discoveries have been made and interpreted. The standard story that Charles Darwin's theory of evolution by means of natural selection was so brilliant that everyone but religious zealots agreed with him is only a caricature of the truth. Darwin's 1859 book proposed more questions than it provided answers, and the scientific endeavor to answer some of those questions has been affected just as much by contingency and chance as the history of life. The places paleontologists looked for fossils and how those fossils have been interpreted have been influenced by politics and culture, reminding us that while there is a reality that science allows us to approach the process of science is a human endeavor.

The following pages tie together the complementary narratives of life's history and our changing understanding of that history. Walking whales, amphibious elephants, feathered dinosaurs, land-dwelling fish, mammals that listened with their jaws, multi-toed horses, and upright apes will be presented through the eyes of the scientists who puzzled over their origins, culminating in what we now understand about the evolution of such creatures. The perspective these stories provide has changed how we interpret the past, and leads us to question some of our most cherished beliefs about our place in the universe.

The Living Rock

Beautiful is what we see,
More beautiful is what we understand,
Most beautiful is what we do not comprehend.
—NICOLAUS STENO, 1673

In October 1666 a French fishing boat spotted an enormous fish off the shores of Tuscany. It was a great white shark, a toothy monster over a ton in weight, and it was quickly caught and hauled onto the beach. It did not perish easily; just when it seemed to die at last it would suddenly begin thrashing the sand with its crescent-shaped tail. So it was tied to a tree to prevent it from flopping itself back into the sea and escaping.

Sharks were common in the Mediterranean, but such a prodigious fish was a rare catch. Word of its capture quickly reached the Grand Duke Ferdinando II at the Medici palace in Florence. Nothing would please him more than to have his cadre of anatomists dissect the *lamia*, but he would have to act quickly. The shark had already begun to rot and its bulk was too great to convey overland to the anatomical theater. There was no option but to cut off the shark's battered and bloody head, which was loaded onto a cart bound for Florence.

The question was who would get the honor of dissecting the rare treasure from the sea. Ferdinando had cultivated a garden of scholars able to do the job, but he chose a young naturalist from Denmark named Nicolaus Steno. At twenty-eight Steno was already known for his keen observations and preternatural skill with a scalpel. As soon as the great fish head arrived Steno prepared to study it.

The dissection was not a quiet affair hidden away in a silent laboratory. The autopsy was a public event, as much a piece of performance art as a scientific deconstruction of a putrid fish. Layer by layer, Steno peeled back the strata of tissues to reveal the primitive inner workings of the sea monster, but the most significant part of the head was among the most obvious: the teeth. Even though Steno made no precise measurements of them, the teeth of this shark would remain on his mind long after he left the anatomical theater.

For as long as anyone could remember, dark, triangle-shaped stones of varying size had been found throughout the European countryside. No one knew what they were, but there was certainly something special about them. Some scholars, like the classical natural historian Pliny, thought they fell from the sky on moonless nights, and others believed they were the petrified tongues of snakes, called *glossopetrae*. One of the most popular beliefs was that they were products of the rock trying to imitate life. The earth, through some kind of plastic force, was reproducing bits and pieces of the living world, or in the words of Steno's contemporary Athanasius Kircher, had been produced by a "lapidifying virtue diffused through the whole body of the geocosm." Regardless of their origin, though, it was widely agreed that such unique stones held special powers. They were often prescribed (in powdered form) as a cure for a variety of common ailments.

The sixteenth-century French naturalist Guillaume Rondelet had a different view. The tongue-stones were curiously similar to the teeth of sharks he commonly saw in the fish markets. Perhaps, as Rondelet publicly speculated in 1554, the *glossopetrae* were not of supernatural origin after all, but had once been arrayed in the mouths of living sharks. This sounded too fantastic to be true, especially since so many of the "shark teeth" were found far inland. It was easier to accept that they were supernatural in origin, and this view still prevailed when the Italian lawyer Fabio Colonna considered the "tongue stones" in 1616. Colonna had been driven to investigate the *glossopetrae* in detail due to the claims of their medicinal value, but after dissecting them for himself he felt safe in concluding that "nobody is so stupid that he will not affirm at once at the first insight that the teeth are of the nature of bones, not stones."

As with Rondelet's conclusion, however, Colonna's findings did not gain traction. What they had proposed grated against "common sense." If an alternate explanation of the origins of the teeth was to catch on it would have to make sense not only of the nature of the teeth but also how they found their way onto dry land.

Steno, working fifty years later, was sure that the *glossopetrae* were true teeth like those from the mouth of the shark. He had seen similar stone teeth in the collection of his former professor, Thomas Bartholin, but he ran into the same questions that Rondelet and Colonna faced. Where had the teeth come from? How had teeth turned to stone? And, for that matter, why were the teeth often found among seashells many miles from the nearest shoreline?

LAMIAE PISCIS CAPVT

EIVSDEM LAMIAE DENTES

FIGURE 5 – A woodcut of the head of a great white shark, as figured in Steno's report on its dissection. The resemblance between its teeth and "tongue stones" found around the countryside started Steno thinking about Deep Time.

As Steno studied the curious petrifactions, he had a startling thought. The fossil teeth had clearly once belonged to living animals, so perhaps the teeth had been shed and deposited in the soft mud of an ancient seabed. They were eventually buried, along with shells and other detritus, by sand and mud that hardened into rock as the years went by. This process entombed the organic productions and eventually transformed them to resemble the rock in which they were found. What had happened to the ancient sea was another matter.

For Steno there seemed to be only two possibilities: either the ocean had once covered the land or the seafloor had become land. The Bible clearly stated that the earth had twice been entirely covered in water, during the Creation and the later Flood, but observations showed that land once under the sea could be pushed up in the aftermath of earthquakes. Both scenarios could have caused the fossils to come to rest in the countryside.

Steno published these early findings in his 1667 report on the shark dissection, *Canis carcharias dissectum caput*, but his work on the subject was only just beginning. The puzzle of the fossils lured him further away from the anatomical theater and more often into the mountains in search of fossil remnants of the ancient seabed. This was a period of emotional agony for him, for he was wrestling with his faith and ultimately converted to Catholicism, but Steno kept up his work and was blindsided by a finding so obvious it was strange that no one had reported on it before.

As Steno ambled the hills around Florence in search of fossils he often saw layers of rock stacked one on top of the other like the pages of a great stone book. This had been appreciated by a few earlier naturalists, Leonardo da Vinci among them, but their findings were either lost, kept secret, or forgotten. Just as with the shark teeth, it would be

Steno who would have to not only rediscover what others had already found, but explain the phenomenon as well.

Steno presented his thoughts on both strata and the fossils in his 1669 essay *Prodromus to a Dissertation on Solids Naturally Enclosed in Solids*. It was only an abstract of a fuller work that he was planning, but an essential part was Steno's idea of how fossils formed. Steno wrote:

> Given a substance endowed with a certain shape, and produced according to the laws of nature, to find in the substance itself clues disclosing the place and manner of its production.

It was not necessary to appeal to "plastic forces" or supernatural phenomena; if you want to understand nature you have to turn to nature itself. Fossils were relics of lost chapters in earth's history that contained clues as to how they were made. Neither Steno nor anyone else could visit the time when Florence was sunk beneath the shark-patrolled waters, but by studying fossils Steno could discover what had happened during the incredibly remote period when the fresh teeth fell to the seafloor.

The layers of rock Steno observed were essential to proving the antiquity of the fossil teeth and shells he had collected. The fossils were found in lower layers of rock, and Steno reasoned that as sediment accumulated on the bottom of the primeval sea the layers began to stack up. The youngest layers would be on top and the oldest on the bottom. Tracing the characteristics of the layers from top to bottom allowed one to travel through time. The belief that the rocks had simply been made that way from the beginning could be cast out. The earth was literally wrapped in its own history.

Frustratingly, though, the geological sections Steno observed were not always arranged in neat, layer-cake sections. Some of them were tilted, which critics argued was inconsistent with the whole idea of stratification since the sea itself could not tilt. But Steno knew this was a foolish argument. Strata were always deposited parallel to the surface of the sea in a horizontal manner. Any flipping, dipping, folding, or tilting occurred long after the deposition of the rock. If strata could be tilted, or even flipped, though, how could you tell which side was up? The solution came from the makeup of each layer. Elements in the layer would naturally settle by weight, with the heaviest at the bottom; any subslice of a layer that contained the heaviest parts was the bottom, and the lightest elements would be at the top.

Strata did not just represent isolated patches. Instead they were con-

tinuous formations of rock that stretched out to the side until they were
stopped by some other formation or material. In a valley the strata on
one side would correspond in the same sequence to those on the other,
despite the space between them. If you knew the arrangement of strata,
you could successively trace them across such gaps.

Together these principles provided a basic guide that could be used
to map the past, but not everyone was ready for this new interpretation
of earth history. There were still some who doubted that fossils were
representations of living creatures. Many fossil shells resembled those
that could be found at the seashore, but there were others that were
clearly different from any modern form. Surely God would not let any
part of the Creation disappear, and the notion that life may have
changed through the ages was strictly against accepted theology. The
Bible could not be controverted, no matter how reasonable Steno's conclusions otherwise seemed.

Steno had promised to work out such difficulties in a full dissertation on his discoveries, but that book was never written. Science did
not allow Steno to reach the absolute truths he yearned for, and after
becoming disaffected with science he became a priest in the Catholic
Church in 1675.

With Steno's voice withdrawn, those who believed that fossils were
spontaneously generated in the bowels of the earth were left to domi-
nate. Though the Englishman Robert Hooke also came to the conclusion that fossils had once been parts of living things after comparing
fossilized wood to wood charred in a fire, there was still widespread
doubt as to how fossils were made and for what purpose. Despite this
reticence, however, it was now reasonable to scientifically discuss the
origins of fossils. Naturalists might not have been inclined to accept
Steno's work, but the history of the earth and life could no longer be
taken for granted.

It was especially important to resolve such issues because it was becoming clear that geology and Scripture did not exactly align. If the
strata reflected the earth's history they told a story that was much more
complex, and theistic geologists strained to keep the Bible and the rock
record in accord. In a scheme presented by English naturalist John
Woodward in his 1695 book *An Essay toward a Natural History of
the Earth*, for example, the Flood was an extraordinarily violent event
in which the suspension of the laws of gravity by God caused the entire
world, organisms and all, to be blended into a slurry. The earth's rock
layers formed out of the murky suspension and fossils were organized
by specific gravity with the heaviest on the bottom. Yet this is not what

geologists in the field saw. The succession of fossil types across strata could not be understood as being organized by weight or size, and appeals to a single (or rare) catastrophe did not account for the formation of strata or fossils.

The present would be crucial to unlocking the past. Even on a timescale as short as a human life it was obvious that the appearance of the coastline changed gradually, with shells and sediment being deposited at a very slow rate. Within the fossil shells scattered about the countryside, naturalists could see the same patterns of growth that shells of living invertebrates exhibited. By the beginning of the eighteenth century the idea that nature was in the habit of throwing up perfect stony facsimiles seemed silly. From the natural origin of the fossils to the formation of the strata they were encased in, it seemed that nature worked by way of familiar, imperceptibly slow processes. The narrow, literalist interpretation of Genesis was breaking apart, and a more allegorical approach was needed to reconcile the testimony of the rocks with scripture.

But the earth did not always speak in a clear voice. Fossils and strata recorded earth's history, but just how far back that history extended was unknown. On the basis of Abrahamic religion it was assumed that the world had a definite beginning and was traveling toward an end, but what if the world was infinitely old? This was the question asked by the eighteenth-century Scottish geologist James Hutton.

Hutton was an eccentric polymath who did most of his work during the intellectual revival known as the Scottish Enlightenment, which took place in and near Edinburgh. Although initially trained as a physician, he was interested in a variety of sciences, and around 1753 he turned to geology. The stones that littered his farm inspired him to ask where they had come from and, in turn, how the formations underneath the countryside had formed.

Hutton mulled over his geological speculations for decades, and it was not until 1785 that he presented an abstract of his theory of the earth to the Royal Society of Edinburgh. In Hutton's view, the land as we know it was cobbled together from the disassociated parts of an earlier world that had been assembled according to natural laws still in operation, and the oceanic abyss was the birthing ground of our present continents. During prehistory, when plants and animals (represented by fossils) lived on the land, sediment and bits of rock fell to the bottom of the sea and hardened. Over time these layers were thrust up, becoming land, and even now there were new future continents forming on the bottom of the sea. The earth was not decaying from a "perfect" state but could restore itself through natural mechanisms.

This made it nearly impossible to calculate how old the earth might be. The world was not struck by catastrophes that suddenly changed its face, but changed slowly and incrementally through processes that could still be observed. Further research would confirm Hutton's ideas of deposition and uplift, but in 1788 he could only state that the strata of the world showed "no vestige of a beginning, no prospect of an end." He would follow these remarks with a full explication of his ideas in the *Theory of the Earth* published in 1795.

Two factors impeded the acceptance of Hutton's view. The first was that his prose was so dense that almost no one could stand to read it. It would only be much later, when his work was summarized by his friend John Playfair, that Hutton's ideas would gain wider acceptance. Second, Hutton's cyclical hypothesis required too much time to be crammed into the traditional Creation narrative, and such historical repetitiveness questioned the linear unfolding of God's work so many Christians believed in. While religious concerns were not as much of an issue in the academic centers of France and Germany, countries at the forefront of science at the time, in England the corroboration of science and scripture was still a major concern. It would be a quirk of translation that would provide succor to those who desired such a reconciliation; it would also provide a target for those who would later carry on Hutton's legacy.

During the late eighteenth century one of the most celebrated French scientists was the anatomist Georges Cuvier. Though of humble social origins, Cuvier quickly impressed his family and teachers at a young age with his ability to retain massive amounts of information about history and nature. He attended lectures and studied science when he could, but it was a chance meeting with a naturalist fleeing the post-French Revolutionary Terror that jumpstarted the young naturalist's scientific career. Scientists

FIGURE 6 – A portrait of Georges Cuvier painted in 1798 by Mathieu-Ignace van Bree.

were not exempt from the violent purge, and among those who took refuge in rural districts was Henri- Alexandre Tessier, a physician and expert on agriculture who delivered lectures anonymously in the town of Valmont.

When the young Cuvier heard these lectures he was struck by their similarity to the works of Tessier, and Cuvier called the naturalist by name. Tessier was shocked. He feared that if his identity became known he would face execution, but Cuvier assured him that he would not reveal the naturalist's identity. This confidence was returned by Tessier, and when the flow of blood ebbed in 1794 Tessier introduced Cuvier to other Parisian naturalists. He fit in perfectly. By 1795 the twenty-six-year-old Cuvier had landed a position as an assistant to the comparative anatomist Jean-Claude Mertrud at the prestigious Jardin des Plantes and was almost immediately elected to the newly formed Academy of Sciences.

Employing the scientific methods developed by colleagues such as the venerable Louis-Jean-Marie Daubenton, Cuvier married geology to comparative anatomy to clear away much of the speculation that surrounded fossil animals. In 1796 he delivered two papers that would frame the debate about fossil animals ever after. The first, on several kinds of fossil elephants found in North America and Siberia, demonstrated that the species of the past were distinct from those living now. What had been suspected by some naturalists was now a reality; species could become extinct. This was backed up by his description of a skeleton transported to Spain from Buenos Aires, Argentina, that Cuvier called *Megatherium*. It was an immense sloth, unlike any living creature, and further proof that the strata of the earth contained the ruins of a lost world that had been destroyed.

As the study of fossils became systematized more resolution could be achieved. It had been recognized by some naturalists that certain kinds of fossils only appeared in particular rock formations. Thus, fossils,

FIGURE 7 – The skeleton of the Buenos Aires *Megatherium*.

to a broad degree, could be used to demarcate successive stages of earth history. Invertebrate fossils were the best for this task, as they were so plentiful and the same types generally persisted for relatively short intervals in the geological column. William Smith, an English geologist who was becoming acquainted with geology during the same time, noticed this trend while working as a surveyor, and traced specific fossils, such as species of the shelled cephalopods called ammonites, across the English countryside.

At the time, however, it could not be understood just how old these formations were. Their relative age could be determined by the succession of fossils, but their absolute age was a mystery.[9] Hence Smith's work focused on the correspondence between fossils and the strata that contained them, and this helped him to produce brilliant geological maps of England. Unfortunately, however, Smith did not have an academic sponsor to speak for him when his work was plagiarized, and this made it so difficult for Smith to make a living that he was eventually cast into the debtor's prison at King's Bench. It was not until 1819 that he was able to continue work as a surveyor in an attempt to eke out a living, and it would take another decade before his work would be recognized by his geological colleagues.

Like Smith, Cuvier also recognized the close association between fossils and divisions of time. In 1808, he and his peer Alexandre Brongniart published a report called "Essay on the Mineral Geography of the Paris Region," in which they observed that the order of the strata surrounding the city could be understood by the distribution of fossils within the rock. The transitions were a little messy, specific forms were not locked just into one layer alone, but certain species showed spikes of prevalence only to dwindle in the next successive layer before disappearing entirely. This method worked over and over again for the identification of geological time and thus meant that fossils could be reliably used to not only differentiate between vast periods of ancient time but to separate one thin slice of earth history from another.

On the basis of invertebrate fossils contained in the surrounding strata, Cuvier and Brongniart were able to determine that the area around Paris had been subject to alternating inundation by seawater and freshwater. Changes occurred cyclically but Cuvier considered them to have a more punctuated pattern than the steady state world Hutton proposed. The sea had rushed in, remained for a time, then quickly receded and was replaced by freshwater before the cycle started again. Such rapid change would surely destabilize environments and cause extinctions, so this regular alternation of catastrophic events could explain the various fossil accumulations, cordoned off by barriers of extinction, which Cuvier saw.

Cuvier most explicitly outlined the role of catastrophes during the earth's history in his 1812 book *Ossemens fossiles*, a synthesis of the work he had begun two decades earlier. In the "Preliminary Discourse" that prefaced the work, Cuvier pontificated on the pattern of ruin and reform over time, and he identified the most recent catastrophe as occurring about 6,000 years ago. Cuvier derived this figure from looking at archaeological and historical evidence. In 1799 the French expedition that recovered the Rosetta Stone also found the mummified remains of birds and other animals in the tombs of Egypt. When carefully examined, these animals—though thousands of years old—were revealed to be identical to their living counterparts, confirming that the last great upheaval had happened before they lived. On the basis of such evidence, Cuvier concluded:

> In this revolution the countries in which men and the species of animals now best known previously lived, sank and disappeared; that conversely it laid dry the bed of the previous sea, and made it into the countries that are now inhabited; that since that revolution the small number of individuals spared by it have spread out and reproduced on the land newly laid dry; and that consequently it is only since that time that our societies have resumed progressive course, that they have formed institutions, erected monuments, collected facts of nature, and combined them into scientific systems.

Simply put, some parts of the previous world sank below the sea while today's continents were thrust out of the water in almost an instant, and the few survivors of the last geological revolution repopulated the planet. New life did not spring up in the wake of extinction, but life was already present and migrated into new areas.

This was not a biblically based argument. Though Cuvier was a Protestant, he kept religion out of his scientific work. Even so, his discussion of a catastrophe, most likely caused by an encroaching sea during the last 6,000 years, was all too easily co-opted by those in England who required that geology conform to religious strictures.

Shortly after it was published in French, the "Preliminary Discourse" was translated into English under the title *Essay on the Theory of the Earth* by Robert Jameson, a Scottish naturalist who subscribed to the idea known as Neptunism which held that the earth's rocks had precipitated out of oceans.[10] The biblical Flood fit well with this geological view, and through a new introduction, end notes, and several revised editions Jameson asserted that Cuvier's work supported the reality of a Deluge that had wiped the world clean during recent history. Jameson's English translation introduced readers to Cuvier as what we would currently call a young earth creationist.[11]

But Jameson's co-option of Cuvier's "Preliminary Discourse" was not the only resource drawn upon to marry the science of geology and Christian theology. William Buckland was a Church of England reverend who was a popular lecturer in geology and paleontology at the time Cuvier's introduction was translated into English. He had a deep love and respect for scripture, yet he was not exactly a literalist. He was perfectly comfortable with the idea that the world's age was inestimably longer than the 6,000 years of biblical chronologies but he was steadfast on the historicity of Biblical events. Buckland believed that there certainly had been a Flood, but it would have been so short that almost no sign of it could be seen in the geological column.[12]

Still, Buckland was willing to change his ideas in the face of new evidence. In 1822 he began to investigate the fossil bones of Kirkdale Cave in Yorkshire. Many of those bones bore tooth marks and were scattered among a collection of chalk-white coprolites, or fossilized feces—sure signs that the cave had once been a hyena den. To confirm that cave hyenas were responsible for this damage Buckland devised a simple, but ingenious, experiment. Working with their keeper, he gave bones to the hyenas of a local zoo and compared the marks the living hyenas made with the ones of the ancient bones. Through the comparison of the fossil bones with the freshly chewed bones and the collection of prehistoric poop, Buckland was able to conclusively demonstrate that the hyenas had actually lived inside the cave during some ancient period and had not just been washed in by the Deluge.

It was a simple discovery, but during a ceremony awarding Buckland the Copely medal for his work, the president of the Royal Society, Humphry Davy, said it presented "a point fixed from which our researches may be pursued through the immensity of ages." Encouraged by his peers, Buckland continued his research into the world before the Flood. He presented these findings in his 1823 work *Reliquiae diluvianae, or, Observations on the Organic Remains Attesting the Action of a Universal Deluge.*

In "Relics of the Deluge," Buckland strained to find balance between two different sets of expectations. The idea that the Deluge had done most of the work in forming the present world was no longer taken seriously by most naturalists, yet Buckland believed that there was evidence for such an event. Caves filled with mammal bones were like time

FIGURE 8 – In a cartoon drawn by his friend Henry de la Beche, a young William Buckland uses the light of science to unravel the mysteries of Kirkdale Cave.

capsules that preserved a record of life before the catastrophe. Once the makeup of this lost fauna was identified their remains could be traced across the countryside, and often these bones were found in deposits of gravel, sand, and clay—materials that were easily transported by water. The bones of hippos, rhinos, lions, hyenas, and elephants in these "superficial deposits" attested to a recent inundation, Buckland believed, which replaced the tropical climate with a milder one.

The weakness of Buckland's hypothesis, however, was that he had presumed that all the cave strata and gravel deposits he examined were of the same age and therefore recorded a single event. As these deposits were probed further it was discovered that they had been formed at different times, and despite his book's popularity its conclusions were contested.[13] The geological pattern did not appear to follow the hurry-up-and-wait cycle of stability punctuated by disasters, and it was at this time that Hutton's view of a cyclical, orderly earth was coming out of hibernation.

The geological revolutions of naturalists such as Cuvier and Buckland required geological events so devastating as to be almost supernatural. Everyone knew of floods, but no one had ever recorded one continent becoming inundated while another rose from the sea. Faced with this dilemma, some geologists began to return to Hutton's view of ongoing, gradual change.[14] The Scottish geologist Charles Lyell, through his three-volume series *Principles of Geology*, would become the champion of this idea and transform the discipline.

FIGURE 9 – A portrait of Charles Lyell, drawn later in his life.

Lyell proposed that while the earth was constantly in flux it was also in balance. A river might erode rock and transport it to the sea, but this destruction was counteracted by volcanoes spewing new rock onto the surface. Every geological process had its counterpart, and the planet was constantly reforming itself in a slow and steady fashion. In the introduction to chapter five of the first volume of *Principles of Geology*, published in 1830, Lyell wrote:

The first observers conceived that the monuments which the geologist endeavours to decipher, relate to a period when the physical constitution of the earth differed entirely from the present, and that, even after the creation of living beings, there have been causes in action distinct in kind or degree from those now forming part of the economy of nature. These views have been gradually modified, and some of them entirely abandoned in proportion as observations have been multiplied, and the signs of former mutations more skilfully interpreted. Many appearances, which for a long time were regarded as indicating mysterious and extraordinary agency, are finally recognized as the necessary result of the laws now governing the material world; and the discovery of this unlooked for conformity has induced some geologists to infer that there has never been any interruption to the same uniform order of physical events. The same assemblage of general causes, they conceive, may have been sufficient to produce, by their various combinations, the endless diversity of effects, of which the shell of the earth has preserved the memorials, and, consistently with these principles, the recurrence of analogous changes is expected by them in time to come.

Even the terrifying earthquake was simple a mechanism by which the earth's uniformity could be maintained. "This cause, so often the source of death and terror to the inhabitants of the globe, which visits, in succession, every zone, and fills the earth with monuments of ruin and disorder," Lyell wrote, "is, nevertheless, a conservative principle in the highest degree, and, above all others, essential to the stability of the system."

But Lyell knew that he not only had to show why he was right but why the opposition, naturalists who favored Cuvier's "revolutions," were wrong. There was no reason to think that geological phenomena had been more severe in the past than now, and Jameson's biblically tinged translation of Cuvier's work allowed Lyell to cast those who favored changes caused by catastrophes as fundamentalists who read scripture too narrowly. Through a rereading of evidence and force of argument, Lyell fostered the view of earth history Hutton had conceived three decades before.

As the earth changed, however, so did life. Paleontologists had clearly shown that the life of the past was different from that of the present, and if the movements of planets and the shifting of the earth proceeded by natural laws then the formation of life should be no different. Hutton himself had considered this in his immense, three-volume monograph of 1794 called *An Investigation of the Principles of Knowledge*. He wrote:

> If an organised body is not in the situation and circumstances best adapted to its sustenance and propagation, then, in conceiving an indefinite variety among the individuals of that species, we must be assured, that, on the one hand, those which depart most from the best adapted constitution, will be most liable to perish, while, on the other hand, those organised bodies, which most approach to the best constitution for the present circumstances, will be best adapted to continue, in preserving themselves and multiplying the individuals of their race.

The culling of organisms ill-adapted to their local conditions would cause species to become altered over the course of generations—but Hutton did not ascribe much power to this mechanism. Despite the Creator's benevolence in allowing organisms to become adapted to shifting conditions there was no reason to think that these changes ever lead to the origin of new species. Some variation and change was expected, but all according to a set of natural laws that maintained the created order.

Naturalists on the continent, and specifically in France, had been willing to go a bit further than Hutton. There were most certainly orderly natural laws that affected biological entities, but they could cause those forms to break the boundaries that were thought to limit what a species might become. The most prominent exponent of this view was Seven Years War veteran Jean-Baptiste Lamarck.

Lamarck had not initially set out to be a naturalist. His family had a strong military tradition, and at the age of sixteen he joined the French

FIGURE 10 – Jean-Baptiste Lamarck in his later years.

army to battle Prussia. Despite his age he was quickly commissioned as an officer, but when one of his fellow soldiers lifted Lamarck up by his head during horseplay it caused the young officer's lymphatic glands to become inflamed. Lamarck had to go to Paris for treatment, and afterward he was given a quiet, but not very well-paid, post in Monaco. Then, after a failed bout studying medicine, Lamarck turned his attention to botany. Both on his own and under the tutelage of the French naturalist Bernard de Jussieu, Lamarck familiarized himself with the flora of France, culminating in his 1778 book *Flora française*.

This work drew the attention of Paris's established naturalists, and by 1781 Lamarck was named the Royal Botanist at the Jardin du Roi (which was renamed the Jardin des Plantes in 1790). But Lamarck's interests were not restricted to plants. He trained himself as a specialist in invertebrates, too, and during the closing years of the eighteenth century he began to study the fossil shells found around Paris.

The shells were different from, but still similar to, modern forms he was familiar with, and Lamarck believed that it would not take much to turn one form into another. Yet he went even further than this. In 1800 he presented a lecture on invertebrates at the Muséum national d'Histoire naturelle in which he proposed that all forms of life had been derived from earlier, simpler forms. Organisms achieved this through their interactions with the environment, hence producing the close fits between animals and the habitats they were found in. The webbed feet of some waterfowl had been produced in such a way:

> The bird which necessity drives to the water to find there the prey needed for its subsistence separates the toes of its feet when it wishes to strike the water and move on its surface. The skin, which unites these toes at their base, contracts in this way the habit of extending itself. Thus in time the broad membranes which connect the toes of ducks, geese, etc., are formed in the way indicated.

It was the habits of the animals that resulted in their form, not the form that dictated function, and these changes gained over the life of an individual were transmitted to the offspring of that animal. Yet as Lamarck continued to consider evolution, this acquisition of traits through habit or willpower became only the mechanism for a greater trend Lamarck saw in the history of life. In his most comprehensive presentation of his views of evolution, his 1809 book *Philosophie zoologique*, Lamarck pointed out that life appeared to be progressive, and the simplest life was always being pushed up a ladder of complexity. As new simple life forms spontaneously arose the mechanism of acquired characteristics came into play, and the habits of the organisms determined their progress up the evolutionary ladder; today's monads could be tomorrow's humans. New life, and hence potential new forms, were always being formed, and if extinction happened at all it was rare.

Lamarck's evolutionary hypothesis was not very well received. Lamarck's contemporary Georges Cuvier was staunchly opposed to it, as it ran against the more systematized view of biology Cuvier had struggled to build. Whereas Lamarck thought form was dictated by function and extinction almost never happened, Cuvier asserted that function and form were tied together in a static package, and the evidence was clear that creatures did go extinct. If they did not, then what had the *Megatherium* and mammoth transformed into? Lamarck seems to have been aware of this difficulty—as he stated that extinction might happen among large, slow-breeding vertebrates—but in general he denied extinction for a vision of nature in which life was continuously moving from lower to higher without being extinguished.

The tension between Lamarck's and Cuvier's views was not so much about evolution as conflicting philosophies about how nature could be understood. In England, however, Lamarck's view was seen as a materialistic proposal that directly contradicted the belief that divine wisdom was manifested in nature. The idea of a progression of organisms from more primitive to more advanced types was especially troubling, but in applying geological principles to living creatures Lyell thought he had found a way out. Clearly organisms were well matched to the environments in which they existed, but if the earth was in a constant, gradual state of flux then those habitats would not be permanent. As the geology of the planet changed, so would environments and the creatures in them, and given enough time the world might rearrange itself into a state approximating that of the age of the dinosaurs. During such a time, Lyell proposed, it would be expected that extinct creatures would reappear to reprise the roles they had fulfilled so long before:

We might expect, therefore, in the summer of the "great year," which we are now considering, that there would be a great predominance of tree-ferns and plants allied to palms and arborescent grasses in the isles of the wide ocean, while the dicotyledonous plants and other forms now most common in temperate regions would almost disappear from the earth. Then might those genera of animals return, of which the memorials are preserved in the ancient rocks of our continents. The huge iguanodon might reappear in the woods, and the ichthyosaur in the sea, while the pterodactyle might flit again through umbrageous groves of tree-ferns.

FIGURE 11 – Henry de la Beche's cartoon of the "Awful Changes" that would occur if, according to Lyell's ideas about the unfolding of the history of life, ichthyosaurs returned to the world during some future time.

This attempt to obscure any notion of progression in the fossil record seemed so absurd that Lyell's colleague, geologist Henry de la Beche, lampooned it in a widely circulated cartoon. Entitled "Awful Changes—Man Found Only In A Fossil State.—Reappearance of Ichthyosauri," the scene depicted "Professor Ichthyosaurus" astride a podium apprising his class of a fossil human skull: "*A lecture*—'You will at once perceive,' continued Professor Ichthyosaurus, 'that the skull before us belonged to some of the lower order of animals; the teeth are very insignificant, the power of the jaws trifling, and altogether it seems wonderful how the creature could have procured food.'" Even as the idea of evolution remained controversial, Lyell's cyclical hypothesis for the pattern of life's history could not mask the very real succession of fossil creatures paleontologists were beginning to uncover.

There had to be some other explanation for this pattern in the rocks, some kind of natural law that regulated the appearance of organisms through time. Over the previous several hundred years naturalists had, bit by bit, substituted the direct actions of a deity with natural mechanisms (albeit ones put in place at the moment of Creation). A literalistic interpretation of Genesis was not sufficient to gain true knowledge of nature, but naturalists feared what they might find when considering the "species problem." What was true for the origins of other animals would be true for our kind, too, and the prevailing religious sentiments of the day made naturalists uneasy about this concept. By the time Lyell published the first volume of *Principles of Geology* such a mechanism was still wanting, but it would be one of his own pupils who would supply it.

Moving Mountains

The theory of the transmutation of species . . . has met with some degree of favour from many naturalists, from their desire to dispense, as far as possible, with the repeated intervention of a First Cause, as often as geological monuments attest the successive appearance of new races of animals and plants, and the extinction of those pre-existing. But, independently of a predisposition to account, if possible, for a series of changes in the organic world, by the regular action of secondary causes, we have seen that many perplexing difficulties present themselves to one who attempts to establish the nature and the reality of the specific character.
—CHARLES LYELL, *Principles of Geology*, Volume 2, 1832

In the evening hours of January 19, 1836, at the height of the Australian summer, a young naturalist reclined on the bank of a sun-warmed river near Wallerang to ponder the weird and wonderful creatures of the great island continent. Earlier in the day he had taken part in a kangaroo hunt, and that afternoon he had seen one of the strangest of all Australia's creatures, a duck-billed platypus, both of which were entirely unlike the animals he was familiar with from Europe. Reflecting on such animals, he could not help but think of Australia as a place created separately from the rest of the world. Later that night he wrote in his journal:

An unbeliever in everything beyond his own reason, might exclaim "Surely two distinct Creators must have been [at] work; their object however has been the same & certainly the end in each case is complete."

But such a claim was obviously absurd, and the presence of a more familiar creature underlined the unity of design in nature. Lying in wait in a little sand pit near the riverbank was an antlion, a vicious insect larvae that sprays passing ants with sand until they tumble down the walls of the pit into its waiting jaws, and in it the naturalist again saw the imprimatur of the Creator:

Without a doubt this predacious Larva belongs to the same genus, but to a different species from the Europaean one.—Now what would the Disbeliever say to this? Would any two workmen ever hit on so beautiful, so simple & yet so artificial a contrivance? It cannot be thought so.—The one hand has surely worked throughout the universe. A Geologist perhaps would suggest, that the periods of Creation have been distinct & remote the one from the other; that the Creator rested in his labor.

This was the young Charles Darwin, pondering the handiwork of the Creator toward the end of his five-year journey around the world on the HMS *Beagle.*[15] At twenty-seven, Darwin was a recent Cambridge graduate just one step away from becoming an Anglican parson, and he was an adherent of the English brand of natural theology that saw the will of God in all of nature. All he could ask for when he got home would be a quiet little country parish in which he could deliver sermons on Sunday and build his natural history collection during the rest of the week.

Enlightened appreciation of nature had long been important to Christianity. During the second century there appeared a popular text in Greek called the *Physiologus* in which nature confirmed the raw power and creativity of God while serving as an allegory for the teachings more explicitly stated in the Bible.[16] The trick was preventing curiosity about the natural world from turning into something pagan. Myth and reality were often bound up together, especially in representation of what existed in far off lands, and during the Middle Ages it became fashionable to store up collections of the weird and wonderful productions of nature. Cataloging and collecting became an activity of empire justified by the need to fully appreciate the Creation, and eventually fairy tales gave way to a natural theology that relied more on the interpretation of nature's aesthetics. By the time Darwin sailed around the world in the early nineteenth century there was nothing untoward in a priest's being preoccupied with collecting beetles, flowers, or fossils. Such activities only supplemented intellectual lessons about God's work.

Given that Darwin's father was a physician and his brother had already taken the same path, however, medicine had been his initial choice of profession. Darwin had cut his teeth as an assistant to his father during house calls in 1825, when he was 16 years old, and in the fall of that year he started his formal training at the University of Edinburgh. He quickly learned that he was not cut out for medicine. The demonstrations in the operating chamber before the advent of anesthesia were horrible spectacles of gore and violence that he could not bear. He remained at the school until 1827 before his father grew frustrated with his lack of progress and made him reappraise his options.

But Darwin's time at Edinburgh had not been wasted. Not only had he learned the finer points of taxidermy from a freed slave, John Edmonstone, but he also became heavily involved in the natural history clubs there. In fact, it was in the school's Plinian Society that he presented one of his first scientific discoveries: that what had previously been thought to be some kind of spore in oysters were in fact the eggs of a skate leech. During this time Darwin also studied under the Scottish naturalist Robert Edmund Grant, an advocate of Jean-Baptiste Lamarck's view of evolution and the transmutationist speculations presented by Darwin's grandfather, Erasmus, in verse. As Grant and Darwin walked to the seashore together the elder naturalist sang the praises of the French transformists and pondered how all life had arisen from a simple common ancestor. (Hence, Grant's preoccupation with sponges, regarded as among the most primitive of organisms.)

Darwin was an attentive student of Grant's, assisting the elder naturalist with research and practicing French so that he, too, could understand what authorities on the continent were talking about, but he was more in awe of Grant than a direct adherent to his views. That species were mutable was still a highly controversial idea. Naturalists were free to consider the heretical notion on their own time, but they could expect fierce public opposition from the social conservatives that controlled the country (as Grant himself most certainly did).

But despite Darwin's aptitude as a budding naturalist he still had to choose an occupation that would gain him the respectability that could never be achieved by searching tide pools and stuffing birds. Since he had neither interest, nor the temperament, for becoming a soldier or a lawyer, this left the Church of England as a relatively safe, easy, and honorable way for him to find his place in society. Although he came from a liberal Whig family full of Unitarians, and would be entering a profession dominated by social conservatives, as long as he did his duty on the Lord's Day he could hunt birds and study sea slugs as much as he liked during the rest of the week. His habits, a liability in the pursuit of almost any other career, would not be a hindrance in the Anglican Church.

Since Darwin's cousin, William Darwin Fox, was already at Christ's College at Cambridge, it was decided that he should do the same. (At the very least, there would be someone there to make sure he was not spending too much time away from his studies.) He started afresh there in early 1828, and though Darwin did prefer his hobbies, especially his new avocation of beetle collecting, he did not wash out of Cambridge as he had Edinburgh. In early 1831, as planned, he passed his final exams and attained his Bachelor of Arts degree. The question was what was to come next, and Darwin would have to wait in Cambridge until June for graduation. He used this time to simultaneously feed his interest in nature and prepare for taking the holy orders.

The Anglican theologian William Paley's *Evidences of Christianity* had been required reading for the finals, and Darwin was so struck by it that picked up Paley's *Natural Theology* during his spring break. No other book could have suited Darwin better. In *The Evidences of Christianity* Paley was primarily interested in standard proofs of God such as miracles; in *Natural Theology* he turned his attention to nature. According to Paley the whole of nature was fine-tuned and harmonious, and each intricately designed part worked with the others to achieve a designed purpose. He summarized this view at the outset in what would become known as the watchmaker argument:

> In crossing a heath, suppose I pitched my foot against a stone, and were asked how the *stone* came to be there, I might possibly answer, that, for any thing I knew to the contrary, it had lain there for ever: nor would it perhaps be very easy to shew the absurdity of this answer. But suppose I had found a *watch* upon the ground, and it should be enquired how the watch happened to be in that place, I should hardly think of the answer which I had before given, that, for any thing I knew, the watch might have always been there. Yet why should not this answer serve for the watch, as well as for the stone? Why is it not as admissible in the second case, as in the first? For this reason, and for no other, viz. that, when we come to inspect the watch, we perceive (what we could not discover in the stone) that its several parts are framed and put together for a purpose, e.g. that they are so formed and adjusted as to produce motion, and that motion so regulated as to point out the hour of the day; that, if the several parts had been differently shaped from what they are, of a different size from what they are, or placed after any other manner, or in any other order, than that in which they are placed, either no motion at all would have been carried on in the machine, or none which would have answered the use, that is now served by it.

Clearly the watch had been designed, its many parts working together for one purpose, whereas the stone was just some uninteresting pebble. (I can only imagine what a geologist would say to this latter point.) Applied to nature, Paley's argument was similar to that which Georges Cuvier had made about the stability of species. Organisms had to be organized in a precise fashion, and any change to those anatomical plans would cause a disequilibrium in which the organism would break down. Whereas Cuvier demurred on how these forms were made, however, Paley had no qualms about attributing their creation to God. The intricate nature of living things attested to what Christians took by faith.

Darwin was delighted by this.[17] Paley's view of a benevolent, carefully designed nature perfectly married his Cambridge studies with his ambitions to be a clergyman naturalist like his teacher John Henslow. Henslow, a botanist at the college, was an enthusiastic supporter of students who took an interest in natural history, and Darwin's zeal for the subject matter made them close friends. Just as with Grant, Darwin supplemented the formal parts of his study by discussing natural history on long strolls with Henslow, and did so often enough that Darwin became known as the "man who walks with Henslow." During the months that he was required to stay at Cambridge Darwin saw Henslow frequently, and through his combination of clerical and natural knowledge Henslow seemed like the perfect person for Darwin to emulate.

But Darwin was not ready to jump into the service of the church just yet. Among the other books he read while waiting for graduation day was the *Personal Narrative* of the German naturalist Alexander von Humboldt. Despite its dull title, the massive volume was an exciting account of the natural wonders of South America, and it sparked Darwin's interest in the tropics. There was so much to see outside England, and as much fun as scrounging for beetles in the English countryside was, he imagined how much grander it would be to find dozens of new varieties on a far-off continent! He did not have the funding for such a long voyage, but a trip to someplace a little closer to home did not seem out of the question. Tenerife, off the northwest coast of Africa, seemed like the perfect destination. Darwin was able to interest Henslow and three other students in participating, too, and his reticent father agreed to underwrite the trip. (After all, his brother Erasmus toured the European continent when he graduated.) Darwin talked almost incessantly about the trip and could not wait to go.

To get Darwin up to speed before the trip, Henslow introduced him to another cleric-cum-naturalist, Adam Sedgwick, who was an expert on geology. Darwin began attending Sedgwick's lectures that spring and was exhilarated by just how much there was left to discover, and he took as avidly to mapping the geology of the countryside around his family's home as he had to beetle collecting. The expedition to Tenerife could not leave soon enough.

Darwin remained steadfast in his ambition, even as most of the members of the trip backed out, including Henslow, but the unexpected death of the last of the willing adventurers stalled Darwin's plans. It was now apparent that if there was going to be any expedition at all he would have to go alone. Conflicted, Darwin rode home to the family manor, but when he arrived there was a fat envelope for him from Henslow. It was an invitation to an even grander adventure; a potential slot on a survey to South America on the HMS *Beagle*.

The Royal Navy ship the HMS *Beagle* had been built in 1820, but other than taking part in the celebration of the coronation of King George IV it had simply sat at the docks, unused. An expedition launched in 1826 would give it new purpose. South America was open for trade, and knowledge of the continent's waterways was essential for British merchants to compete with other Europeans. The *Beagle* was fixed up and refitted to accompany the HMS *Adventure* on a survey of the waters and coast of southern South America, a vast expanse of mountains and sparse grasslands.

It was not an easy journey. The isolation faced at sea proved to be too much for the captain of the *Beagle*, Pringle Stokes, who shot himself just over two years into the trip. In the wake of this tragedy the ship came under the command of a young officer named Robert FitzRoy. After he took over, FitzRoy spent a great deal of time trying to recover a small boat that had been stolen by some of the native Fuegians and, in his frustration, FitzRoy took several of their children hostage. They were to be educated in England and returned as missionaries, it was decided, and while the official reason for the second voyage of the *Beagle* was to continue the hydrographical survey, an equally important impetus was the return of the abducted children to England.

Given the fate of the *Beagle*'s earlier captain, FitzRoy knew all too well the way the isolation of a long sea voyage could prey on the mind. He cast about for a suitable gentleman companion to accompany him. Leonard Jenyns, a clergyman naturalist, was among the first to be tapped, but he demurred, as did Henslow. Both were already established and had other duties at home. The offer trickled down to Darwin through his connection to Henslow, and he seemed like a perfect choice because he was not married and not yet a member of the clergy. After convincing his father that the trip would be beneficial, on December 27, 1831, twenty-two-year-old Charles Darwin left London for South America and points beyond.

It was not a pleasant trip for Darwin (at least while he was at sea). Lacking a bunk, he slept in a hammock strung high above the floor of his cabin, and he constantly battled with seasickness. Still, Darwin's onshore journeys made the entire ordeal worth while. The wonders of South America were open to him, giving him the chance to make new discoveries in a place most of the naturalists back home had never been.

After reading the first volume of Charles Lyell's *Principles of Geology* the young naturalist was enthusiastic about fieldwork in a distant land. Henslow had warned Darwin not to take Lyell too seriously, as Lyell was proposing a dressed-up version of Hutton's old ideas that did not leave room for great catastrophes to shape the earth, but Darwin drank deeply from the pages of Lyell's work. It made more sense that the rocks of the earth should have been historically shaped by causes now in operation. Even if large scale changes seemed slow, local events could occur very rapidly.

On the morning of February 20, 1832, Darwin was lying down in the woods around the town of Valdivia, Chile, when the ground began to shake. The earthquake was so severe that he could not stand until it was over, and when it abated he ran into the town to see people running about slanted houses that had been upright the night before. The damage was even worse in the city of Concepción, which the *Beagle* reached two weeks later. The city was in shambles, an enormous smoldering pile of rubble.

There was a geological lesson to learn from the humanitarian disaster. On the shore nearby Darwin saw that the mussel beds had been thrown up several feet above the high-tide line. It was at once apparent that the rise of continents above the waves did not happen in a single catastrophic event; they had been pushed by starts and stops over immense periods of time until what had once been the sea floor became mountains. Geological monuments as impressive as the Andes had been created by small changes over an incalculable amount of time, and as Darwin searched the strata of South America he began to find evidence that the procession of life may have also been precipitated by natural processes.

Though not a paleontologist, Darwin was familiar with fossils. His education in geology instructed him that the earth as we knew it was only the latest iteration of a succession of fossil worlds once dominated by strange creatures like icthyosaurs and mammoths. Despite the distinctiveness of creatures in each era, however, there were hints of continuity between organisms that spanned the geological divide.

Darwin knew from reading the books in the *Beagle* library that fossils of creatures strange, yet familiar, were to be found in South America. Previous accounts told of giant armadillo shells, and everyone knew

Cuvier's description of a giant sloth, a gargantuan version of the living arboreal mammals in the continent's jungles. Darwin found similar fossils and began to ship them back to England to be studied by expert anatomists. As he did so he wondered what these animals might be and, in a letter to Henslow sent between October and November 1832, Darwin confided that he believed he had found ancient remains of a kind of rodent, some parts of a giant sloth, and the plates of the curious giant "armadillo" that had been previously reported. These latter fossils, especially, suggested a correspondence between the life of the present and the past. The ancient bits of shell plating were not only armadillo-like in construction, but were often found alone, just as the empty shells of armadillos were more often found than an entire body. This correspondence was even more apparent among the smaller remains of a fossil agouti. In a field notebook entry made in February 1835, Darwin commented that he hoped that the remains of these mammals would "be one more small instance, of at least a relation of certain genera with certain districts of the earth. This co-relation to my mind renders the gradual birth & death of species more probable."

Such a relation was supported by the discovery of another mysterious fossil mammal. In early 1834 Darwin found bones that he initially attributed to a mastodon near Puerto San Julián in Argentina, but, the geological context of the fossil led him to doubt this initial hypothesis. It had been embedded in gravel, not the type of substrate for a lush forest to take root, even in ancient times, and it was more reasonable to think that this animal had lived in a scrubby habitat akin to that favored by the camel-like guanacos Darwin often saw during his journey. Given the essential match between an organism and its environment, it seemed more likely that the "mastodon" was really a creature similar to a guanaco.

The life of the past was clearly related to that of the present, but Darwin had not yet become a transmutationist. While Darwin saw that the "death" of species might have occurred gradually the appearance of new species in the fossil record still seemed sudden, and Darwin had no idea what kind of creative law might account for the dispersal of related forms through time. As his musings on the Australian riverbank in 1836 showed, he was still convinced that animal forms had been contrived through the genius of one Creator.

When the *Beagle* returned to England later that year Darwin was ready to be received as an accomplished naturalist. In addition to helping write up some of the official accounts of the journey, Darwin met with various experts to discover the secrets of the various specimens he had collected.

FIGURE 12 – A portrait of Charles Darwin, painted by George Richmond in the late 1830s.

When it came to fossils there was no naturalist more skilled than Richard Owen, a surgeon who applied his knowledge of anatomy to natural history. Although vehemently opposed to wild theorizing about evolution (the socially conservative Owen had just kicked Darwin's old evolutionist mentor, Robert Grant, out of the Zoological Society), Owen was fascinated by ideas about form and growth being discussed on the continent.

As Owen began to work on the fossils he soon found that Darwin had gathered a greater variety of fossil mammals than even the young naturalist had expected. One of the first to be examined was a skull Darwin had purchased from a rancher for eighteen pence that, in life, had belonged to an unusual hoofed mammal the size of a rhinoceros. Owen named it *Toxodon*, and the naturalists of London were thrilled that Darwin had brought them such a strange beast.

The mystery of the "armadillo" plates was soon resolved, as well. For years naturalists had debated whether the plates were to be attributed to *Megatherium* or an entirely different sort of animal. The fossils Darwin brought back allowed Owen to confirm that they did represent a distinct

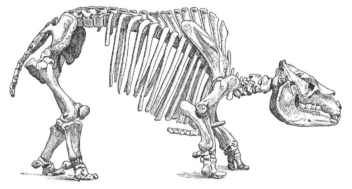

FIGURE 13 – The skeleton of *Toxodon*, the enormous fossil mammal Darwin discovered in Patagonia.

creature, which he named *Glyptodon* in 1839. It was a close relative of the giant sloths, but one encased in a tough shell of osteoderms.

Toxodon and *Glyptodon* were among the fossil celebrities that Lyell presented to the Geological Society in February 1837. (The "guanaco" from San Julián would later turn out to be an entirely new kind of mammal, a hoofed mammal which looked like a trunked camel, which Owen would name *Macrauchenia*.) While Owen focused on their description, Lyell considered their implications, specifically that there was a succession of related types in South America over time. This is what Darwin had suspected since digging them out of the ground, but the question remained: Why did one form succeed another type? Their extinction was a puzzle, but even more confounding was the force that had brought them into existence in the first place.

FIGURE 14 – Restorations of many of the peculiar fossil mammals which lived in South America during the Pleistocene. During his journey to Patagonia, Charles Darwin collected the bones of many of these animals. Clockwise from the top left: *Doedicurus*, *Glyptodon*, *Hippidion*, *Toxodon*, *Mylodon*, *Megatherium*, *Macrauchenia* (dog included for scale).

The birds Darwin and FitzRoy collected during a stop in the Galapagos Islands provided a crucial clue. During his visit to the archipelago Darwin did not think much about the birds. He simply collected them and stuffed them, often not even bothering to record the island from which they had come. Among the few that Darwin took the trouble to label by island were four mockingbirds, all relatively similar to each other. As he began to jot down notes about these specimens in his "Ornithological notes," however, he thought that each type might have been a variety suited to its particular island, not specially made but slightly altered to a different habitat:

> If there is the slightest foundation for these remarks the zoology of Archipelagoes—will be well worth examining; for such facts would undermine the stability of Species.

The popular view at the time was that species radiated outward from a "center of creation." When a new species came into existence, by whatever mechanism, it spread out from this central spot to occupy suitable habitats nearby. If this were true, however, all of the mockingbirds Darwin found should have been the same from island to island. They were not that far from each other, after all, and the only way to describe the differences between them would be to assume that the Creator had called a slightly different bird into existence on each island. That made a mockery of the Almighty, and it was more sensible to think that each of the birds was a slightly modified type derived from an aboriginal species.

When ornithologist John Gould had a look at the specimens, however, he found that they were not just variations. Each mockingbird was a distinct species. This was an even more powerful expression of what Darwin was already starting to think about. Perhaps species were not as immutable as had been supposed.

Discoveries made by other naturalists also fueled Darwin's speculations. In the spring of 1837 two fossil primates, one from India and the other France, were discovered, and Lyell joked that they would provide transformists like Lamarck "many thousand centuries for their tails to wear off, and the transformation to men to take place." Darwin was becoming less willing to dismiss such thoughts with a laugh. His own view was that such species were like individual creatures in that they had a set time of growth, senescence, and death. This would perfectly explain the extinction of species like the *Glyptodon* and *Macrauchenia* he found in South America as there did not seem to be any sign of a catastrophic event that wiped them out. Their numbers simply seemed to dwindle before they disappeared altogether, like the withering stem of a plant. Species appeared to have a lifespan just as individuals did, but the same old problem remained of where these animals came from in the first place.

For a time Darwin thought that the production of "monsters" might provide an answer. Maybe some ancient *Glyptodon* literally gave birth to a more modern-looking armadillo, and so the production of species would result from the birth of individuals vastly different from their parents. This would explain why there was not a series of connecting "links" between living and fossil animals, but this was not a steadfast rule. The numerous species of finch that Gould cataloged contradicted this trend by exhibiting a graded series of closely related types.

The questions surrounding such evolutionary changes enthralled Darwin. Other naturalists shied away from the topic, lest their discoveries bestialize humanity in the process, but Darwin could not stop wondering about what had produced both the living and fossil species he had collected aboard the *Beagle* voyage. Species were not static, that much was clear, but how did they change? Were there limits to transformation? How long did it take and by what natural force could it be achieved? These intellectual obstacles had to be overcome. As a young naturalist looking to solidify his career he could not be too careful.[18] The reception Lamarck and Grant received made it clear that open speculation on such subjects would not be taken lightly. Instead, in the middle of July 1837, Darwin started a new "transmutation notebook" to privately record his thoughts.

Rather than build on his previous musings on new species as mutants, though, Darwin began to consider how circumstances might favor particular forms over others. It was clear that individuals of a species varied, and that variations were tied to the intricacies of environmental change, but the exact nature of the connection was unclear. Perhaps the environment stimulated the birth of new forms, or maybe some other natural force was at work.

The pattern of life's history also demanded an explanation. To Darwin's eye there did not seem to be a linear progression of the "primitive" to the "advanced." If organisms changed according to the conditions of their environment then no such linear form of evolution made sense, and Darwin instead conceived of evolution as producing a tree that was united by common ancestry but cut back by extinction. The extinct mammals of South America were some of the withered limbs. In his notebook Darwin wrote, "We may look at Megatherium, armadillos, and sloths as all offsprings of some still older type some of the branches dying out."

Darwin had now moved beyond merely recognizing the succession of types to positing common ancestry. The living and extinct animals he had collected were all derived from still older creatures, each bearing adaptations to the environment it inhabited. Without a solid mechanism to constrain the growth of the tree, however, these isolated observations and bits of data could just as well be shoehorned into the old views of centers of creation, as Lyell and others had done. Darwin had recognized that adaptation to local conditions would lead to the origin of new species and the extinction of those unable to change, but what drove this process was another question.

Darwin scoured the biological literature to find more evidence for what he was considering, but in late September 1838 he turned to *Essay on the Principle of Population* by Thomas Malthus for a break from his research. It was a book with a bleak message. Human population was growing rapidly, placing an increased strain on available resources, but this growth did not go unchecked. In addition to the struggle for resources, war, famine, disease, and other factors reduced human populations when competition became too severe. Even though the absolute size of the population still grew, the total potential population size was never realized because of the action of these agents.

Darwin was already familiar with this Malthusian view. It had been debated for years in the context of whether the poor should be given handouts or left to their own devices. Yet when he read Malthus's book it struck him in a new way. Organisms in the wild, too, seemed to be able to produce more offspring than would actually survive. There were only so much space and food to go around, and thus organisms were in a constant struggle with each other to make a living. The few best able to acquire these resources would survive and reproduce, while those not as well equipped would perish.

What made the difference in this struggle were inherited variations. Individuals that possessed variations that gave them an advantage in the struggle to survive would pass those traits on to their offspring, so small changes would accumulate to cause the origination of entirely new species. This was a far cry from the beneficent nature that had enchanted Darwin when he read Paley's work. Instead, as counterintuitive as it seemed, the death of many would allow the few to thrive.

Hence Darwin's evolutionary hypothesis was based upon several observations that could not be refuted. First, individuals of any species vary, and these variations are heritable. Second, these variations provided an advantage to some individuals in the competition for resources. Third, these variations made them not only more likely to survive but, more importantly, to leave more offspring with those advantageous traits. Other naturalists, such as James Hutton, had stumbled upon a similar idea before, but they had generally thought it worked within established limits and could not institute change on a grand scale. Darwin thought differently. Given these conditions life *had* to evolve, and this process would leave behind a branching pattern of organisms all connected by common ancestry.[19]

Darwin knew that he would require a mountain of evidence to support his ideas. Even if he constructed a well-argued case some naturalists might still reject his ideas on the grounds that it stripped all power from the Creator. Yet just as Darwin was secretly beginning to identify the rudiments of his evolutionary vision, naturalists were beginning to reject well-worn arguments about divine intervention in the creation of life.

The growing dissatisfaction with the answers provided by natural theology was epitomized by the reaction to the eight-volume series *The Bridgewater Treatises*, published between 1833 and 1840. The series had been commissioned by Francis Henry Egerton, a British naturalist whose deathbed act of contrition was funding a series of books to remind the people that God still spoke to them though nature. Each of the eight parts was written by a different authority, from William Whewell's *Astronomy and General Physics Considered with Reference to Natural Theology* to the final entry in the series, *Chemistry, Meteorology, and the Function of Digestion, Considered with Reference to Natural Theology* by William Prout.

William Buckland, the Oxford geologist who had solved the mystery of Kirkdale Cave, was tapped to write the sixth volume, which appeared in 1836. It was an ode to the benevolent Creator from the perspective of paleontology. In considering the *Megalosaurus*—an immense fossil reptile that Buckland described in 1824 on the basis of a few skeletal fragments and believed to be akin to a monstrous terrestrial crocodile (and, in 1842, deemed to be a dinosaur by Richard Owen)—the naturalist insisted that the recurved fangs of the reptile were signs of God's wisdom.

> In a former chapter I endeavoured to show that the establishment of carnivorous races throughout the animal kingdom tends materially to diminish the aggregate amount of animal suffering. The provision of teeth and jaws, adapted to effect the work of death most speedily, is highly subsidiary to the accomplishment of this desirable end. We act ourselves on

this conviction, under the impulse of pure humanity, when we provide the most efficient instruments to produce the instantaneous, and most easy death, of the innumerable animals that are daily slaughtered for the supply of human food.

If there had to be death and suffering, it was better that death come quickly and efficiently. God was not so cruel as to give the *Megalosaurus* the appetite for flesh and make it such a bumbling predator that tortured its victims to death.

The *Ichthyosaurus* also embodied a divine lesson. A marine reptile known from several complete skeletons, many of which had been found by one of the earliest fossil hunters, Mary Anning, it had paddles similar to those of a whale, a crocodile-like head, vertebrae like those of a fish, and a sternum like that of the enigmatic creature from Australia, the platypus. A scoffer might look at this collection of parts and think that God made a mistake, but Buckland insisted that the fact that *Ichthyosaurus* combined parts from various animals showed that God used the same parts in the designs for His creatures over and over again:

> The introduction to these animals, of such aberrations from the type of their respective orders to accommodate deviations from the usual habits of these orders, exhibits a union of compensative contrivances, so similar in their relations, so identical in their objects, and so perfect in the adaptation of each subordinate part, to the harmony and perfection of the whole; that we cannot but recognise throughout them all, the workings of one and the same eternal principle of Wisdom and Intelligence, presiding from first to last over the total fabric of Creation.

Such arguments generated more derision than agreement. The series, called the "Bilgewater Treatises" by critics, was an anachronism better suited to Paley's time. Buckland's plea seemed more of an effort to hold back liberal threats to conservative orthodoxy than to understand nature as it truly was.

The idea that the birth and death of organisms were regulated by natural laws was becoming more acceptable. This view was formally articulated by Charles Babbage in his unofficial *Ninth Bridgewater Treatise*. It was not enough to look at the pattern of nature and say, "This fossil reptile speaks of God," Babbage argued, but the process by which the Creator's thoughts were translated also required elucidation. For Babbage it seemed that relatively simple laws of mathematics and physics could build up to become components of even grander laws

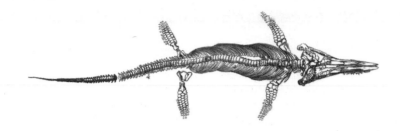

FIGURE 15 – The skeleton of an *Ichthyosaurus.*

that regulated the machinations of the universe. If this was true then it was not unreasonable that the Creator regulated life by similar means.

Babbage took the geological record, both of the rocks themselves and the remains of ancient life contained within them, as a proof of this trend. Strange forms flourished in ages past only to be wiped out and replaced by superior creatures better suited to the changing world. All of this, Babbage proposed, was governed by elegant, simple laws that the Creator imposed upon the matter of the universe, knowing full well from the beginning how history would unfold.

Yet this new vision of change according to natural laws posed new problems. Whatever applied to the rest of nature applied to our species, too, so it could not be argued that our species was the product of a miracle when such intervention had been foresworn from the rest of nature. It was impossible not to recognize our resemblance to apes and monkeys, but to think that we had actually descended from such animals was still enough to make many naturalists shudder. Such fears over the bestialization of humanity made a tract anonymously published in 1844 an instant sensation. Entitled *Vestiges of the Natural History of Creation*, it was both beloved and despised.

Published anonymously by the Scottish publisher Robert Chambers, *Vestiges* complemented Babbage's view of natural change. The fossil record, especially, provided unequivocal evidence that Providence had set laws to regulate evolution, but contrary to Darwin's private notion that evolution was not progressive Chambers believed that life strove ever upward through time, with each stage forming the foundation for the next. Life followed the great hierarchy: fish, amphibians, reptiles, birds, and mammals all appeared according to the traditional order. This was closely tied to changes on the globe, with each group fitted to different environments by biological destiny.

As far as vertebrates were concerned, Chambers believed embryos had a fixed trajectory, and different groups diverged earlier in the process than others depending on the conditions of the environment at the time. If the development along this path was stopped early, a fish was produced; if it was allowed to progress a mammal was formed. The process could be sped up or slowed down, but external factors did not so much change embryos as bring about predestined changes:

> Thus, the production of new forms, as shewn in the pages of the geological record, has never been anything more than a new stage of progress in gestation, an event as simply natural, and attended as little by any circumstances of a wonderful or startling kind, as the silent advance of an ordinary mother from one week to another of her pregnancy. Yet, be it remembered, the whole phenomena are, in another point of view, wonders of the highest kind, for in each of them we have to trace the effect of an Almighty Will which had arranged the whole in such harmony with external physical circumstances, that both were developed upon our planet is but a sample of what has taken place, through the same cause, in all other countless theatres of being which are suspended in space.

Written in a plain manner accessible to almost any adult reader, *Vestiges* was an immediate hit and began running through new editions quickly. Chambers's conclusions were in doubt, but whether or not he was correct was overshadowed by the controversial idea that life was not locked into a static order. The vanguard of traditional religion was quick to respond. A reply published the following year by S. R. Bosanquet opened with a *tsk, tsk* at the public's credulity, likening Chambers's freewheeling discussion to a "wanton and deformed adulteress," that seduced minds that should be focused on God. Others, while not so severe in their rhetoric, made similar charges, and for a time there was a small industry of books denouncing *Vestiges*. The scientific theorizing in the book was bad enough, but among the most vocal critics were those who were offended by Chambers's theology.

Charles Darwin, who had written out a private sketch of his evolutionary theory two years earlier, was both charmed and frustrated by *Vestiges*, especially since some readers attributed it to his hand. No one was quite sure what he was cooking up, but by the time *Vestiges* was published a number of naturalists knew that he was actively considering how life might evolve. He even cultivated a close group of expert friends—among them Charles Lyell, the botanist Joseph Hooker, and Richard Owen's young rival Thomas Henry Huxley—to help him formulate his ideas on the subject.

Hooker broached the subject of the *Vestiges* with Darwin soon after its publication, writing that he was "delighted" by its spirit even if it was scientifically unsound. Darwin replied that he was not nearly as amused, for "[Chambers's] geology strikes me as bad, & his zoology far worse." This made him feel "much flattered & unflattered" at the suggestion that he had written it. Even so, *Vestiges* provided a test whereby Darwin could see how people might react to such ideas. The withering criticisms Chambers's book confirmed that Darwin needed to make a solid, convincing case based upon hard evidence that would be respected by his naturalist peers. He already had the mechanism that would produce the expected pattern of life seen throughout history, but he would have to gather even more data to make the operation of this "secondary law" consistent with life both past and present.

To this end Darwin corresponded with authorities, carried out breeding experiments with fancy pigeons, studied barnacles, and built up the solid base of observation required to support his theory. This would all culminate in what he referred to as the "Big Book," tentatively called *Natural Selection*, which would present the full theory to his colleagues.

A work of such importance could not be rushed, but not everyone would wait up for Darwin to get on with publishing. While Darwin settled down to a quiet life in the country with his wife and family, the tales of his adventures in South America aboard the *Beagle* inspired other younger naturalists. Among these were Henry Walter Bates and Alfred Russel Wallace, two Englishmen who resolved to travel to South America to solve the question of the origin of species.

In 1848, Bates and Wallace sailed for Brazil, and once they arrived they pushed into the jungle along the Amazon River. They worked together at first, but after growing tired of one another's company they split up and continued the endeavor separately. Their respective journeys would be very different. Whereas Bates remained in the Amazon until 1859, by 1852 Wallace had collected enough specimens to sell in England, including living birds, coatis, monkeys, and other vertebrates that Wallace had to pull along with him and feed every day. After an arduous journey back out of the jungle, Wallace made it to a port near the coast and was able to secure passage on a boat home.

A few days later, Wallace watched as a fire that had started onboard consumed the ship, leaving him and his fellow passengers stranded at sea. Save for a few notes, everything was gone. It was ten days before another ship rescued him and the other survivors in the lifeboat.

Once home, Wallace lacked the financial security that had been bequeathed to Darwin and, like other naturalists such as T. H. Huxley, he worked feverishly to pursue his passion. With the exception of a few esteemed professorships doled out by colleges or prestigious societies, there were no official jobs for naturalists. They had to make their own opportunities, either by entering the medical profession or becoming collectors and selling the exotic specimens they found. Wallace was a master at the latter craft, and after a stint of eighteen months (during which he introduced himself to the growing scientific elite of England through several books and papers), Wallace was off again, this time for the South Pacific, to collect more specimens.

Wallace arrived on the Malay Archipelago in 1854, and he would remain there for eight years. This time, though, he sent his findings back home at intervals. Among the most significant was an 1855 paper entitled, "On the Law Which Has Regulated the Introduction of Species." While it did not explain precisely how organisms evolved, Wallace did notice that new species closely resembled those already in existence, meaning that evolutionary steps were built upon what had come before.

This sounded very familiar to some naturalists back in England, particularly Charles Lyell and Edward Blyth, both of whom knew of Darwin's thoughts on the same subject. Blyth, in particular, shot off a hastily written and rambling note about it to Darwin right after the

paper's publication, pointing out the similarities between what Wallace had been proposing and what Darwin had been working on for two decades. Darwin did not take the hint. He thought that Wallace was proposing another variation of progressive creation that was of no threat to his own nascent theory. Yet Wallace's article was enough to get Lyell thinking more seriously about species change, and in 1856 Darwin finally explained his hypothesis to his mentor. Lyell did not agree but knew that if Darwin did not publish soon his primacy would be in danger. Rather than try to hide his theoretical

FIGURE 16 –Alfred Russel Wallace, photographed in Singapore in 1862.

ambitions, however, Darwin let Wallace in on what he had kept from so many other naturalists. On May 1, 1857, Darwin wrote to Wallace (who was still in Indonesia).

> I can plainly see that we have thought much alike & to a certain extent have come to similar conclusions. In regard to the Paper in Annals, I agree to the truth of almost every word of your paper; & I daresay that you will agree with me that it is very rare to find oneself agreeing pretty closely with any theoretical paper; for it is lamentable how each man draws his own different conclusions from the very same fact.—
>
> This summer will make the 20th year (!) since I opened my first-notebook, on the question how & in what way do species & varieties differ from each other.—I am now preparing my work for publication, but I find the subject so very large, that though I have written many chapters, I do not suppose I shall go to press for two years.—

Wallace replied that he was glad that someone was at least interested in his paper (many of those who read it wished he would just get back to collecting and stop theorizing), and that he, too, intended to more fully explicate his hypothesis. Since Wallace planned to remain in the field for three or four more years, Darwin believed that he would have ample time to write up his unique perspective on evolution.

Darwin was not prepared for the shock that hit on June 18, 1858. Enclosed in a letter was an essay Wallace had written that was strikingly similar to what Darwin had been working on for years. Wallace recognized that evolution could be affected by natural selection, and even used some of the same phrases that Darwin employed. Darwin panicked. He immediately wrote to Lyell that his "words have come true with a vengeance that I sh^d. be forestalled," and was in such a frantic state that he rushed off two additional letters before Lyell replied to the first.

Even though Wallace did not present specific instructions to publish the paper, Darwin could think of a no more honorable course. If he did

Figure 17 – Charles Darwin in 1854, just a few years prior to the publication of *On the Origin of Species*.

not pass the paper along he would certainly be accused of suppressing Wallace's work to the advantage of his own. Even worse, Darwin's baby son, Charles, was sick with scarlet fever and passed away on June 28. It seemed like Darwin's life was unraveling even as he approached the exposition of the idea he had labored over for so long.

Darwin's confidants Hooker and Lyell devised a compromise. In a public joint announcement before the Linnean Society, Wallace's essay would be presented with one of Darwin's sketches of his ideas and a few letters; it was all that Darwin could put together on such short notice, even though he had never intended for them to be read publicly.

The ideas presented in the documents were very similar, yet not exactly identical. Darwin, for his part, couched his mechanism in explicitly Malthusian terms:

> Now, can it be doubted, from the struggle each individual has to obtain subsistence, that any minute variation in structure, habits, or instincts, adapting that individual better to the new conditions, would tell upon its vigour and health? In the struggle it would have a better *chance* of surviving; and those of its offspring which inherited the variation, be it ever so slight, would also have a better *chance*. Yearly more are bred than can survive; the smallest grain in the balance, in the long run, must tell on which death shall fall, and which shall survive. Let this work of selection on the one hand, and death on the other, go on for a thousand generations, who will pretend to affirm that it would produce no effect, when we remember what, in a few years, Bakewell effected in cattle, and Western in sheep, by this identical principle of selection?

The analogy to the modification of domesticated animals was i mportant. Everyone knew that a judicious farmer, by determining which individuals of his stock could breed, could create entirely distinct varieties of livestock in a matter of years. If humans could do it, why not nature? Yet this was just one part of what Darwin was proposing. If there really is a struggle for existence we should expect for nature to produce a diverse variety of species such that numerous organisms could all live in the same place.

> The varying offspring of each species will try (only few will succeed) to seize on as many and as diverse places in the economy of nature as possible. Each new variety or species, when formed, will generally take the place of, and thus exterminate its less well-fitted parent. This I believe to be the origin of the classification and affinities of organic beings at all times; for organic beings

always seem to branch and sub-branch like the limbs of a tree from a common trunk, the flourishing and diverging twigs destroying the less vigorous— the dead and lost branches rudely representing extinct genera and families.

Darwin did not imagine evolution as striving to a preconceived end as Chambers and others did, but it could be said that each new species was superior (or at least better adapted to the new conditions) than its ancestor and so would outcompete its progenitor. Thus the origin and death of species were inextricably tied together as life branched out to fill all available spaces.

Wallace, however, started on a different tack. He began by saying that domesticated animals were not a good example of how evolution might work, for it was well known that domesticated animals released into the wild would eventually return to the "wild type." It was only natural selection in the wild that could effect permanent change, but from there Wallace's remarks are eerily similar to Darwin's. Just like Darwin, Wallace saw variations providing advantages to some individuals and not others, and under times of extreme environmental stress the newly formed species would be superior to the parent stock. There were plenty of dead branches in the tree of life. Evolution by natural selection made sense of both the process and pattern of evolution, leaving Wallace to conclude:

> This progression, by minute steps, in various directions, but always checked and balanced by the necessary conditions, subject to which alone existence can be preserved, may, it is believed, be followed out so as to agree with all the phenomena presented by organized beings, their extinction and succession in past ages, and all the extraordinary modifications of form, instinct, and habits which they exhibit.

The Linnean Society was not impressed. Natural selection was a well-known phenomenon and was thought to have no great effect on nature. At best it was a preserving force that caused species to remain stable, and at worst it was a destructive force that was the antithesis of creative power. The naturalist Patrick Matthew, for one, had anticipated what Darwin and Wallace called natural selection in 1831 when he published a passage on the mechanism as a preserving force in the appendix of the obscure volume *On Naval Timber and Arboriculture*. The fact that he buried it in such an out-of-the-way place gives some indication of his estimation of its importance. The head of the Linnean Society, Thomas Bell, had a similar estimation of evolution by natural

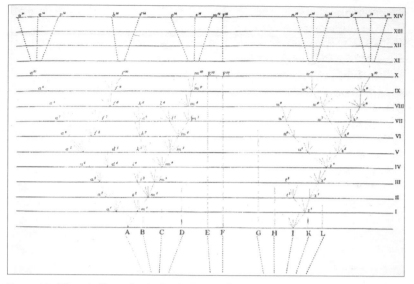

FIGURE 18 – The only illustration in *On the Origin of Species*—a diagram of evolution's branching pattern.

selection. According to him, 1858 passed without a single idea of special importance coming up in the society.

Both Darwin and Wallace were disappointed by the way other scientists shrugged off the idea that both had been working on for so long, but for Darwin, at least, this was something of a blessing. If the idea was not given special attention, neither did it generate the firestorm of controversy he had for so long feared. Now he had to quickly publish an abstract of his full views before someone else scooped him. This abstract, a shadow of what *Natural Selection* was meant to be, was called *On the Origin of Species by Means of Natural Selection*.

Darwin opened his book by recapitulating the development of his ideas. "When on board H.M.S. 'Beagle,' as a naturalist," Darwin explained, "I was much struck with certain facts in the distribution of the inhabitants of South America, and in the geological relations of the present to the past inhabitants of that continent." The "succession of types" Darwin saw firsthand had set him on the trail of "that mystery of mysteries," evolution, but at that time the fossil record presented more problems than solutions.

According to Darwin's theory, evolution was gradual. This did not mean that evolution proceeded at a constant pace, but that it happened in a graded, stepwise fashion. There was a continuity of forms, and if these forms could all be traced they would coalesce into a nested hierarchy of successively older common ancestors. In 1859 the fossil record as a whole was consistent with what evolution predicted, with the appearance of different vertebrate types occurring in a sensible fashion (fish before amphibians, amphibians before reptiles, etc.), but a direct fossil confirmation of what Darwin was proposing was, as yet, unidentified.

Darwin addressed this problem in two chapters: "On the Imperfection of the Geological Record" and "On the Geological Succession of Organic Beings." "Geology assuredly does not reveal any such finely-graduated organic chain," Darwin admitted, "and this, perhaps, is the most obvious and gravest objection which can be urged against my theory." Paleontologists were always working from an incomplete record, of which Darwin wrote:

> For my part, following out Lyell's metaphor, I look at the natural geological record, as a history of the world imperfectly kept, and written in a changing dialect; of this history we possess the last volume alone, relating only to two or three countries. Of this volume, only here and there a short chapter has been preserved; and of each page, only here and there a few lines. Each word of the slowly-changing language, in which the history is supposed to be written, being more or less different in the interrupted succession of chapters, may represent the apparently abruptly changed forms of life, entombed in our consecutive, but widely separated, formations.

Yet Darwin could do more than emphasize the contingencies of the fossil record. As he pointed out, naturalists already recognized that species appeared and disappeared in a piecemeal fashion over time. This constant flux of forms in competition with each other was most beautifully expressed in the last paragraph of the book:

> It is interesting to contemplate an entangled bank, clothed with many plants of many kinds, with birds singing on the bushes, with various insects flitting about, and with worms crawling through the damp earth, and to reflect that these elaborately constructed forms, so different from each other, and dependent on each other in so complex a manner, have all been produced by laws acting around us. These laws, taken in the largest sense, being Growth with Reproduction; Inheritance which is almost implied by reproduction; Variability from the indirect and direct action of the external conditions of life, and from use and disuse; a Ratio of Increase so high as to lead to a Struggle for Life, and as a consequence to Natural Selection, entailing Divergence of Character and the Extinction of less-improved forms. Thus,

from the war of nature, from famine and death, the most exalted object which we are capable of conceiving, namely, the production of the higher animals, directly follows. There is grandeur in this view of life, with its several powers, having been originally breathed into a few forms or into one; and that, whilst this planet has gone cycling on according to the fixed law of gravity, from so simple a beginning endless forms most beautiful and most wonderful have been, and are being, evolved.

When the book went up for sale on November 24, 1859, Darwin was both elated and anxious. That morning he wrote to the ornithologist T. C. Eyton, "My Book will horrify & disgust you," but he hardly could have expected the response of his old geology mentor, Adam Sedgwick. Sedgwick, not holding back, admitted that he read Darwin's book with "more pain than pleasure," continuing, "Parts of it I admired greatly; parts I laughed at till my sides were almost sore; other parts I read with absolute sorrow; because I think them utterly false & grievously mischievous."

This deeply distressed Darwin. Still, it was the opinion of his close friends Hooker, Lyell, and Huxley that mattered most. Darwin almost pleaded for Huxley, especially, to tell him what he thought of the book, and was greatly relieved when Huxley stated that it was the greatest work on natural history he had read in nearly a decade. More than that, Huxley, ever opposed to the hold of the supernatural in the sciences, was ready to defend Darwin's evolutionary theory in public. "I am sharpening up my claws & beak in readiness," Huxley wrote, for he knew that opposition to Darwin's theory would be manifested in more than just personal letters from naturalists of the old guard.

One of the first attacks appeared in the *Edinburgh Review* and, despite being published anonymously, was immediately recognizable by its self-referential style as having been written by Richard Owen. While he praised *On the Origin of Species* for fully considering an evolutionary mechanism, Owen asserted Darwin had not approached the topic of the "mystery of mysteries" with due reverence. *On the Origin of Species* contained a few observational gems, but these tidbits were awash in a sea of speculation. For Owen, Darwin's jottings on the fossil record were the most embarrassing sections of all.

Lasting and fruitful conclusions have, indeed, hitherto been based only on the possession of knowledge; now we are called upon to accept an hypothesis on the plea of want of knowledge. The geological record, it is averred, is so imperfect! But what human record is not? Especially must the record of past organisms be much less perfect than of present ones. We freely admit it. But when Mr. Darwin, in reference to the absence of the intermediate fossil forms required by his hypothesis—and only the zootomical zoologist can approximatively appreciate their immense numbers—the countless hosts of transitional links which, on "natural selection," must certainly have existed at one period or another of the world's history—when Mr. Darwin exclaims what may be, or what may not be, the forms yet forthcoming out of the graveyards of strata, we would reply, that our only ground for prophesying of what may come, is by the analogy of what has come to light.

Even worse, the fossil species that were known seemed to show no change. Extinct creatures such as the *Ichthyosaurus*, *Megatherium*, mastodon, and dinosaurs appeared suddenly in the fossil record, showed no sign of transforming into another type, and were entirely wiped out. Citing himself, Owen ambiguously maintained "that perhaps the most important and significant result of palaeontological research has been the establishment of the axiom of the continuous operation of the ordained becoming of living things."

The Oxford geologist John Phillips also used geology to attack *On the Origin of Species*. In an 1860 volume called *Life on Earth*, Phillips maintained that the fossil record was filled with the rapid appearance of species followed by little change, not the continuous replacement of older forms by new species derived from them. This opposition was as theologically based as it was scientific, and it could have been just as easily written by Paley or Buckland.

Phillips, like Owen and others, was worried that Darwin's evolutionary view undermined our special place in nature. Natural selection was like a dark mirror image of natural theology. Rather than being an expression of divine benevolence, Darwin's "entangled bank" was a warzone in which each generation did battle on a Golgotha made of perished species. This was "Nature, red in tooth and claw" that Tennyson feared, and Darwin's view presented little solace for those that wished to "stretch lame hands of faith . . . And faintly trust the larger hope."

Similar appraisals were made by paleontologists abroad, as well. Like Owen, the German paleontologist Heinrich-Georg Bronn thought Darwin's proposal to be the most coherently argued mechanism for evolution hitherto forwarded, but he could not accept it as the answer to the mystery of the origin of species. The Swiss authority François-Jules Pictet thought similarly, convinced that evolution was a worthwhile question to consider but disagreeing with Darwin's hypothesis that all life was connected by common ancestry. The general outline of what was known of the fossil record suggested that evolution was a reality, but Darwin's explanations in *On the Origin of Species* seemed both too violent and too weak to drive the changes paleontologists observed. Even Huxley and Lyell, among Darwin's staunchest supporters, took issue with their colleague's evolutionary formulation. Lyell was still unsettled by the prospect of human evolution, and Huxley preferred large-scale jumps by mutation and other non-Darwinian hypotheses to make sense of evolution over the whole of earth's history.

This state of affairs deeply frustrated Darwin. In an 1861 letter to the American paleontologist Joseph Leidy, who positively referenced Darwin's views, Darwin wrote, "Most palaeontologists (with some few good exceptions) entirely despise my work; consequently approbation from you has gratified me much.—All the older geologists (with the one exception of Lyell, whom I look at as a host in himself) are even more vehement against the modification of species than are even the palaeontologists."

Even if other naturalists could not accept natural selection, however, Darwin's book, unlike those of Lamarck or Chambers, had made evolution a question that could no longer be ignored. Darwin would continue to investigate and publish on evolution and the major effects of small, cumulative changes for years to come, from the details of animal and plant domestication to orchids and the lives of worms in soil, but determining the fossil record's contribution to our understanding of evolution would largely fall to others.

Oddly enough, such work had already been underway while Darwin was beginning to sketch his ideas on natural selection. In the controversy over Darwin's evolutionary mechanism it was seemingly forgotten that, just a few years prior to the publication of *On the Origin of Species*, several creatures were proposed as embodying a transition of vertebrates from water to the land. This groundwork had been laid by Richard Owen, who in spite of his opposition to Darwin's view of life, provided some of the most compelling evidence for his rival's evolutionary theory.

From Fins to Fingers

"The arm of the Man is the fore-leg of the Beast, the wing of the Bird, and pectoral fin of the Fish."
—RICHARD OWEN, *On the Nature of Limbs*, 1849

It was an ugly fish. Its snub-nosed head topped a sinuous, eel-like body, and in place of fins it had only wispy filaments. Nor did it behave like a proper fish; occasionally it came to the surface to gulp air before returning to the bottom to search for snails and other tasty morsels. It had none of the beauty or charm of its smaller, brightly colored relatives that darted about the same Amazonian rivers and swamps, and its unusual form deeply puzzled Johann Natterer.

Natterer was an Austrian naturalist who accompanied thirteen other explorers to Brazil in 1817. Commissioned in honor of the marriage of the daughter of Austria's Emperor Franz II to the crown prince of Portugal, the expedition was believed by the Austrian nobility to be a perfect opportunity to catalog the natural riches of the tropical Portuguese colony. At that time natural science, pressed into the service of empire, involved shooting, trapping, and pickling as many specimens as possible. The detailed study of the specimens had to wait until scholars back in Europe could pick through the spoils.

By the time Natterer returned to Austria in 1835 he had amassed an impressive collection of tens of thousands of specimens starkly different from the fauna of his native country: jewel-like butterflies, garishly colored frogs, and many other intricately beautiful creatures. But one of the most vexing finds was the ugly brown fish he had pulled from the depths of the Amazon. He brought his specimens to Leopold Fitzinger, an expert on reptiles at the Imperial Museum in Vienna.

Natterer and Fitzinger's dissection of the preserved specimen only made things more confusing. As they opened it up and picked among its olive-green bones they found that, in addition to the expected gills,

FIGURE 19 – A caricature of Richard Owen "riding his hobby" (a *Megatherium*).

the creature possessed rudimentary lungs. What was a fish doing with the organs of a "higher" class of animals? Though Natterer believed the new species to be an aberrant fish, Fitzinger thought it was a particularly piscine amphibian because of its lungs. The fish threatened to breach one of the great divisions in nature, and to underscore its tenuous place between worlds it was given the name *Lepidosiren paradoxus*, the "paradoxical, scaled salamander."

At about the same time that Natterer and Fitzinger were scrutinizing the anatomy of their Amazonian lungfish, Richard Owen received a similar specimen from Africa. It had been found wrapped in a ball of mud and mucus about a foot below the scorched African soil in Gambia, a small nation on the great "ear" of the elephant-shaped continent. Having no idea that his Austrian peers had described a similar fish, Owen issued a brief description, calling it *Protopterus anguilliform*, in honor of its eel-like appearance.

When news of Natterer's *Lepidosiren* reached him, Owen was struck by how similar it was to his *Protopterus*, which he reassessed accordingly. (Some less charitable authorities even suggested that Natterer's collection from Brazil had gotten mixed up with one from Africa. That there were two such fish on opposite sides of the Atlantic was almost too fantastic to believe.) In 1841 Owen announced that both the African and South American forms belonged to the same genus. The lungfish from the Amazon would keep its name while *Protopterus* was renamed *Lepidosiren annectens*, the "connecting, scaled salamander."[20] Arriving at this seemingly simple conclusion had been no easy task:

> It may truly be said that since the discovery of the *Ornithorhynchus paradoxus* [the duck-billed platypus], there has not been submitted to naturalists an animal which proves more forcibly than the *Lepidosiren* the necessity of a knowledge of its whole organization, both external and internal, in order to arrive at a correct view of its real nature and affinities.

Superficially, both species looked like fish, complete with scales and rudimentary fins, but their internal anatomy complicated the matter. In addition to their simple lungs, other organs of the *Lepidosiren*, such as aspects of the circulatory system, resembled their counterparts in amphibians. Owen was faced with Natterer and Fitzinger's dilemma: was *Lepidosiren* an amphibianlike fish or a fishlike amphibian?

In most fish, the nasal openings are used for smell and do not play any role in respiration, and Owen noted that both species of *Lepidosiren* had nasal openings that did not connect to their lungs. Though Owen would later be shown to be incorrect on this point, the overall constellation of "fish traits" that he cataloged supported his decision that *Lepidosiren* was a fish.

Owen's decision to call *Lepidosiren* a fish has traditionally been taken as a signal of his distaste for all things evolutionary. A staunch opponent of Charles Darwin's view of evolution, Owen often injected awkward, pious prose into his scientific papers. This led later generations of naturalists to brand him a villainous creationist who did everything in his power to stifle evolutionary ideas.

Owen would have been shocked by this modern caricature, just as he was mortified by Darwin's assertion that he was a defender of the religiously charged concept that species were fixed entities which could never change. In *On the Origin of Species* Darwin wrote that Owen was among the eminent experts of paleontology who ferociously maintained the "immutability of species." Owen's reaction to this was recorded in a letter Darwin sent to his friend and mentor, Charles Lyell, less than a month after Darwin's book was published:

> I have [had a] very long interview with Owen, which perhaps you would like to hear about, **but please repeat nothing.** Under garb of great civility, he was inclined to be most bitter & sneering against me. Yet I infer from several expressions, that at bottom he goes immense way with us.—He was quite savage & crimson at my having put his name with defenders of immutability.

Owen had a right to be cross with Darwin. While Darwin was still privately mulling over that "mystery of mysteries" Owen was doing the same in public. The species name he chose for the African lungfish, *Lepidosiren annectens*, revealed his transformist speculations.

At the time *Lepidosiren* was discovered, naturalists recognized several distinct vertebrate groups: fish, reptiles (including amphibians), birds, and mammals. Thanks to the intellectual baggage carried over from the God-ordained Great Chain of Being each succeeding group was generally thought to be "higher" than the one before it, with humans crowning all organisms. The seemingly clear boundaries between these divisions were muddied by the *Lepidosiren.*

When Owen examined the preserved corpses of the lungfish he found that they possessed a mix of archaic and advanced traits. It was as if they had been made of spare parts. In the formation of its gills, for example, Owen wrote that the *Lepidosiren* exemplified "a most interesting and hitherto unexampled transitional structure" between bony and cartilaginous fish while its fins, on the other hand, resembled the embryonic limbs of amphibians. In this way an adult *Lepidosiren* could be thought of as an overgrown embryo from a "higher" class of vertebrate, a true transitional form between fish and amphibians.

The anatomical pattern Owen saw in his *Lepidosiren annectens* was consistent with what he had seen elsewhere in his studies. All vertebrate skeletons appeared to have been built on a general body plan, an anatomical blueprint, from which any form could be derived. To see this pattern, however, the generalized blueprint had to be distinguished from products created from it.

Looking at the diversity of forms in nature it quickly becomes apparent that many different organisms share similar anatomical structures. The pinching claws of a crab, for example, superficially resemble our own arms, in that the crab appears to have shoulders, elbows, wrists, and hands. Yet a crab's "arm" consists of a soft interior surrounded by a tough, chitinous exoskeleton; there are no bones in it. It is an analogous structure, or a case of convergence, in which the crab's appendage is similar in form and function as our own arms but made of entirely different parts.

Homology, the possession of similar traits shared due to ancestry, is the counterpoint to anatomical analogy. The arms of a bat, a frog, a human, and a crocodile are all slightly different modifications of one shared anatomical construction. There is an upper-arm bone (the humerus), articulated with two lower-arm bones (the radius and ulna) at the elbow, connected to the wrist where the bones of the hand, including the fingers, are placed. The arms may have different functions but they are all modified versions of the same basic arm structure; thus, the humerus in our arms is homologus with those in the upper arms of a crocodile, our fingers are homologus with those a bat uses to support its wings, and so on. Owen underscored this connection with a beautiful comparison inspired by a statue of the goddess Victory about to slay a bull in the frontispiece to his 1849 essay *On the Nature of Limbs.* Owen imagined the bones inside both the angel and the sacrifice and numbered them to show the homolgous limb bones in both.

Owen did not stop at pointing out the common pattern underlying the form of vertebrate arms and legs. Through homology he attempted to work backwards to find the least common denominator of the vertebrate form through which any vertebrate skeleton could be created with only slight modifications. This vertebrate archetype looked like little more than a bony tube, but in it Owen saw the glimmerings of greatness manifest in the human skeleton.

What Owen needed now was a mechanism to drive the transformations. Considerations of "secondary laws" were already replacing previously held beliefs about creation by divine fiat, but rather than look to external forces, as Darwin had, Owen preferred internal drives. Owen envisioned the production of vertebrate types as the result of a competition between two natural forces pulling against each other. One force caused bodies to repeat certain parts over and over again, like the segments of an earthworm or vertebrae that compose our spinal column. This force dominated in the bodies of organisms traditionally thought of as being simple in construction like insects and worms. The second competing force, which generated more diversity, was considered by Owen to be an adaptive force that modified segments into new shapes. Its action was most commonly seen in "higher" or more complex animals, and the most superior organisms were those that could best overcome the influence of the repetitious force.

These speculations put Owen at odds with some of his allies. As a social conservative, he depended upon the support of those who feared the concept of evolution. That life could evolve was a revolutionary idea that threatened to throw the created order, and hence the social order, into chaos, a fear made real by the French Revolution some years before. As expressed in the children's hymn that Cecil Alexander would write a few years later in 1848,

> All things bright and beautiful,
> All creatures great and small,
> All things wise and wonderful,
> The Lord God made them all.
>
> Each little flower that opens,
> Each little bird that sings,
> He made their glowing colours,
> He made their tiny wings.

> The rich man in his castle,
> The poor man at his gate,
> God made them high and lowly,
> And ordered their estate.

As it was in nature, so it was in society, and to destabilize the order of nature was to deny the divine order of the universe. Owen attempted to find a middle way to express the operation of natural laws that would not offend his social allies, and in the conclusion to *On the Nature of Limbs* he wrote:

> To what natural laws or secondary causes the orderly succession and progression of such organic phaenomena may have been committed we are as yet ignorant. But if, without derogation of the Divine power, we may conceive the existence of such ministers, and personify them by the term "Nature," we learn from the past history of our globe that she has advanced with slow and stately steps, guided by the archetypal light, amidst the wreck of worlds, from the first embodiment of the Vertebrate idea under its old Ichthyic vestment, until it became arrayed in the glorious garb of the Human form.

Evolutionists were perplexed by Owen's fuzzy mix of theology and science, while the anatomist's social allies were horrified that Owen had suggested that nature was not immutable. By attempting a compromise Owen pleased no one, though his ideas were influential.

Contrary to his later opinion of natural selection, Owen's formulation of the vertebrate archetype gave Darwin an anatomical base from which the fixity of species could be attacked. Through homology, even a species as glorified as our own could be connected to the lowly, mud-grubbing lungfish. Bone for bone, our skeletons formed according to the same basic anatomical framework. Owen had detected the pattern, but Darwin had the mechanism, and in the margin of his copy of *On the Nature of Limbs* Darwin scribbled:

> I look at Owen's Archetypes as more than ideal, as a real representation as far as the most consummate skill and loftiest generalization can represent the parent form of the Vertebrata.

For Darwin, Owen's archetype was a hypothetical common ancestor for all vertebrates, and it was natural selection, the interaction between organisms themselves, that could turn the bony tube that was Owen's archetype into the panoply of vertebrate species. It would be left to the paleontologists to fill out the details of just how these changes had occurred. Embryology and anatomy could help naturalists identify these shared structures, but the record of transformation was to be found among fossil species.

Once again it was Owen who began to fill the gap. In 1847 the German paleontologist Georg August Goldfuss described the remains of a "primeval reptile" that he called *Archegosaurus*. It was similar to a crocodile, with a long snout full of sharp teeth fit for snatching small prey from the water, yet it in some ways it more closely resembled amphibians than reptiles. The crenulations inside its teeth, for example, placed it among the labyrinthodonts, extinct amphibians that looked like giant salamanders. When Owen inspected the bones, though, he saw that the skull more closely resembled its counterpart in *Lepidosiren* than those of reptiles. Much like the lungfish, it seemed to contain a mix of characters, and just two months before the publication of *On the Origin of Species*, Owen stood before the British Association for the Advancement of Science and said:

> The *Lepidosiren* and *Archegosaurus* are intermediate gradations, one having more of the piscine, the other more of the reptilian, characters. The *Archegosaurus* conducts the march of development from the ganoid fishes to the labyrinthodont type, the *Lepidosiren* to the perennibranchiate type.

This was a bold statement for the time. Here Owen was identifying two transitional types that marked a great divergence in the "march of development" of vertebrates. *Archegosaurus*, on the one side, was transitional between fish like gars (ganoids) and early amphibians, while *Lepidosiren* represented the type from which certain salamanders that retained gills throughout life (perrenibranchiates) had been derived. Even though, as late as 1841, Owen had insisted that there was a natural barrier between fish and amphibians that could not be crossed, he believed that together these creatures broke down "the line of demarcation between the Fishes and the Reptiles."

Despite the weight this evidence could have afforded this theory, however, Darwin was cautious in identi-

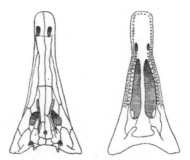

FIGURE 20 – The skull of *Archegosaurus*, an early amphibian thought by Richard Owen to be relevant to the origin of land-dwelling vertebrates.

fying specific species, living or fossil, as the transitional forms that his hypothesis predicted. In *On the Origin of Species*, Darwin acknowledged that *Lepidosiren* was a connecting form between living groups, but more importantly it was a powerful example of how natural selection operated. For Darwin, the *Lepidosiren* showed that if organisms became isolated from the intense competition in the "struggle for existence" they might persist, virtually unchanged, for extended periods of time. Hence *Lepidosiren* was like a "thin straggling branch springing from a fork low down in a tree." Its role as a "persistent" form was more important than its potential spot as a transitional form.

The finely graded series of fossils documenting the evolution of fish to the first amphibious vertebrates was still missing. While it initially seemed that a fish akin to *Lepidosiren* had been gradually modified into something like to *Archegosaurus*, this evolutionary trajectory had a serious problem. The evolution of the first terrestrial vertebrates required the evolution of limbs, a change that the lungfish was in a poor position to initiate.

For all their other amphibianlike traits, lungfish did not possess anything even approximating a primordial limb. All they had were small, bone-filled tendrils that were difficult to envision as being the forerunners of arms and legs. Naturalists debated whether lungfish or other forms were the type from which tetrapods, the first four-legged and land-dwelling vertebrates, had been derived, and it was not until the latter part of the nineteenth century that a good candidate was found. Described by J. F. Whitleaves in 1881, the fossil fish *Eusthenopteron* had a branching series of bones that would have been enclosed in flesh and supported its pectoral fin. It had a single bone connecting the fin to the body, akin to our humerus, and in the group of smaller bones that followed anatomists could see the building blocks for our own limbs. This arrangement seemed adequate for an early fish that was going to venture out onto land and would require stout limbs to support itself.

With fish like *Eusthenopteron* as a morphological starting point, anatomists began to envision what the hypothetical transitional stages between its limb and those of the earliest tetrapods would have looked like. Frustratingly, however, year after year rolled by without the discovery of any fossils that would test these ideas. In his 1922 study *The Origin and Evolution of the Human Dentition*, the American anatomist W. K. Gregory was forced to admit that even though *Eusthenopteron* and its relatives could be thought of as "standing relatively near to the ancestors of the Tetrapoda . . . the connecting links are lacking from the geological record." Just as resistant to solution was the functional question of why fish had hauled themselves out of the water.

That early tetrapods evolved from fish was without a doubt. The fossil record clearly showed that the Devonian world, in which vertebrates only lived in the sea and now dated between 416 to 359 million years ago, had been followed by a time of rapid radiation of early terrestrial vertebrates. What could have effected such a massive change?

In his 1917 textbook *Organic Evolution*, the American paleontologist Richard Swan Lull reviewed the competing hypotheses. Could the fleshy-finned fish have crawled onto land to escape predators? No, for the fish would still be amphibious and have to raise their young in the water, thus leaving the offspring vulnerable even if the parents were safe. Perhaps the lure of food on land enticed the fish to crawl onto the muddy banks of the Devonian swamps? It sounded plausible, but food in the water was just as abundant, if not more so, than on the shores. Nor did it seem that the "lure" of oxygen in the air could have triggered the change paleontologists knew had to have taken place.

The only idea that Lull felt stood up to scrutiny was that the end of the Devonian was marked by intense fluctuations in climate, a notion that had been proposed the previous year by the geologist Joseph Barrell in a lecture before the Geological Society of America. For Barrell, the ancient deposits of the Late Devonian, often containing rocks the color of dried blood, did not indicate any kind of watery paradise for tetrapods and their ancestors. Instead, he took them to represent a dry, semi-arid climate similar to that of modern equatorial Africa, where there were dramatic differences between the wet and dry seasons.

The kind of climate in which Owen's *Protopterus* lived provided a good analog for the conditions endured by the Devonian ancestors of tetrapods. Fish flourished during the rainy season in equatorial Africa, but with the onset of the dry season rivers, lakes, and streams disappeared, leaving behind nothing but cracked earth and fish skeletons. It would have been the same in the prehistoric past, and the fact that many Devonian fish skeletons were found together, Barrell argued, meant that they had been crowded together in shrinking, stinking ponds in just such a manner.

In such harsh circumstances most fish would perish, but Owen's lungfish showed that survival was possible even if escape was not. As bodies of water turned into miasmas of toxic, oxygen-depleted muck, lungfish were able to draw oxygen from the air via their lungs. Then, when all moisture was depleted, they went into a kind of hibernation beneath the soil until the rains returned to release them from their earthen cocoons. The fishy ancestors of the first tetrapods likely had the same defense against drying out.

Eventually, however, there came a time when the rains did not return quickly enough. If the fleshy-finned fish trapped in the isolated ponds were to survive they would have to set off across the baking mudflats in search of more water sources. They would have propped themselves up on their fins and dragged their bodies over the searing mud, and those that could wriggle the farthest would survive to perpetuate that strength in the next generation. After this cycle had been repeated enough times, limbs would have carried the first amphibious creatures further than any of their fish ancestors. It was this struggle for existence that set the stage for the later pageant of tetrapod evolution. After the first tetrapods evolved, vertebrates could exploit all that the land had to offer; we, too, owed our own existence to the success of the intrepid Devonian fish.

Unfortunately there were no fossils to document this change. The oldest known trace of a tetrapod, an enigmatic track named *Thinopus antiquus*, suggested that by the end of the Devonian limbs and fingers had already evolved. (Though there has long been doubt as to whether the ambiguous impression is a track at all.) The Drying Pond Hypothesis was the most popular explanation for what occurred during the undocumented part of the fossil record, and it would be most closely associated with one of the most influential vertebrate paleontologists of the twentieth century, Alfred Sherwood Romer.

Romer worked during a time when evolutionary science was undergoing a major change. From the last decades of the nineteenth century into the 1930s many paleontologists rejected Darwin's idea that natural selection was the primary driving force behind evolutionary transformations. Darwin's vision predicted orderly, graded steps between forms, while many paleontologists saw that some species appeared abruptly in the fossil record and others persisted virtually unchanged for millions of years. Hence they preferred to believe that evolution was guided by internal forces that pushed organisms toward particular endpoints. (Even Barrell, in proposing what sounded like a Darwinian scenario in his hypothesis for the origin of tetrapods, preferred not to say whether it was natural selection or some other evolutionary force that had modified the desperate fish.) These ideas fell broadly under the banner of orthogenesis, the hypothesis that evolution was striving toward particular goals, but no one could say how these alternate mechanisms worked.

Romer was among the naturalists who began to reestablish the importance of natural selection to evolutionary science. Taught by

W. K. Gregory, who was more sympathetic to natural selection than many of his peers, Romer took a keen interest in the relationship between form and function in the vertebrate body, something best understood in terms of Darwinian mechanisms and not internal driving forces. Along with his colleague George Gaylord Simpson, Romer helped to reconcile the fossil record with what Darwin had predicted, and their findings were combined with those of geneticists and population biologists to form the Modern Evolutionary Synthesis. By the 1950s the sloppy orthogenetic ideas that had so pervaded paleontology were all but eradicated.

The Drying Pond Hypothesis, with its emphasis on the evolution of new forms to perform the novel function of walking on land, fit neatly within Romer's view of evolution. What many paleontologists called the "invasion of the land" was not a hopeful step toward "higher" forms of vertebrate life but a dire risk that only seemed glorious in hindsight. And the evolution of limbs was only part of the story, Romer argued. It was not until the erstwhile fish started to consume arthropods (which had made the transition to terrestriality long before vertebrates did) that the true explosion of tetrapod forms began. The basic equipment to move on land had probably evolved by the close of the Devonian, but the impetus to use it did not come until there was something on land for early tetrapods to go trundling after.

Alternatives to this adaptive story were still proposed from time to time, but they could not compete with the romantic imagery of a lowly fish struggling against the elements. In popular restorations *Eusthenopteron*, the fleshy-finned fish from the Devonian of Canada, was often placed with its tail in the water and its forefins on the muddy bank, literally spanning the divide between the aquatic and terrestrial realms. Yet the next phase of the evolutionary series remained elusive. A fish akin to *Eusthenopteron* must have initiated the changes that culminated in early amphibians such as *Eryops*, a salamander on steroids from the 295-million-year-old rock of Texas, but these forms were just bookends to the evolutionary series. The osteological volumes that could be slotted between them had yet to make their full public debut.

The first signs of fossil creatures that would fill the evolutionary gap were not found in the accessible deposits of Europe and North America, but in a much harsher landscape. A series of failed attempts to reach the North Pole by Swedish explorers in the 1890s led to the chance discovery of Devonian-age fish fossils in the icy rock of Greenland. Vestiges of warm, primordial pools were locked in the frozen rock in the shadow of the Celsius Berg, but the petrified scraps were not enough to merit any expeditions to recover more fossils.

As the years rolled by, however, Greenland attracted international attention for its possession of a more useful kind of fossil treasure: oil. Crude petroleum was becoming the lifeblood of developing nations, and in 1929 Norway and Denmark were racing to map and control Greenland's resources. Once again, science was pressed into empire's service, and those who best understood the geology of Greenland stood a better chance of controlling the natural riches contained within its boundaries.

Thankfully some of the geologists sent to Greenland were curious enough to do more than just look for signs of oil. As they mapped the island's stratigraphy some noticed the same kinds of fossilized fish the Swedish explorers had found decades before. One of the geologists employed by the Danish expedition of 1929, a Swede named O. Kulling, even found some fishlike scales that did not quite match any prehistoric fish then known.

When Kulling's colleague Erik Stensiö examined the fossils he determined that they had come from some hitherto unknown vertebrate, but more fossils would be needed to figure out what it was. As explorations of Greenland continued the Swedish geologists continued to pick up fossils, and in 1931 the leader of that year's expedition, Gunnar Säve-Söderbergh, found skull fragments from a creature unlike any of the fish common in the Devonian deposits. It appeared to be the earliest tetrapod ever discovered, and Säve-Söderbergh named it *Ichthyostega*.

Subsequent forays into the field brought back even more parts of *Ichthyostega*. Little by little, the body of the tetrapod was pieced together, but before a full description could be carried out Säve-Söderbergh succumbed to tuberculosis in 1948. The analysis of the fossils ground to a halt. Even though the bones of this odd creature had been recovered, its significance to the origins of the first tetrapods was still largely unknown.

Ichthyostega was too important to be left collecting dust, however, and the responsibility of describing it was passed on to Erik Jarvik, one of Säve-Söderbergh's assistants who had accompanied him into the field starting in 1932. Jarvik seemed to work at a glacial pace, issuing reports on partial aspects of the skeleton every now and then, leaving the anatomy of *Ichthyostega* mostly a mystery. There was little that could be done about this. To scoop Jarvik by finding more fossils of the same animal and describing them first would breach the unspoken code of conduct among fossil hunters. The scientific community would have to wait. (A full report on the creature would not be published until 1996.)

Yet even at an early date it did not appear that *Ichthyostega* was all that paleontologists had been hoping for. In 1956 Jarvik published skeletal and fleshed-out restorations of what *Ichthyostega* would have looked like, and in some ways it seemed to be too specialized to be the ancestor of all later tetrapods. It was not so much an intermediate form between fish and amphibians, Romer (among others) thought, as a creature close to the common ancestry of reptiles and amphibians. It represented the oldest skeletal evidence of a tetrapod, finally supplanting the relatively uninformative track that Lull cited, but it already had a neck, shoulders, arms, legs, hands, feet, fingers, and toes. The fossil gap between fish and tetrapod seemed as wide as ever.

By the 1980s the situation was becoming intolerable. Paleontologists interested in the origin of tetrapods either had to specialize in the array of fleshy-finned fish that came before them or the various amphibious vertebrates that succeeded *Ichthyostega*. There was nothing in between to study, or at least nothing not already spoken for by other scientists. This presented a substantial problem for the English paleontologist Jenny Clack.

After struggling to find a niche for herself in academia as an expert on early tetrapods, Clack sought out a new direction in which to take her work. At this point Jarvik was still sitting on *Ichthyostega*, and a new early tetrapod that would be named *Tulerpeton* in 1984 was out of reach, as it was being described by Soviet scientists. It seemed like the only hope of finding something new was to go back to Greenland.

Without a good lead on a new fossil locality, launching an expedition to Greenland was too much of a gamble. Even if the harsh physical conditions could be surmounted Clack was doubtful that the Swedish paleontologists who had worked under Säve-Söderbergh and Jarvik had left anything worth picking up. The old dig sites had likely been cleaned out.

Clack's hunt for another fossil locality of the right age and type led her to Cambridge geologist Peter Friend, who had done fieldwork in Greenland during the 1960s and 70s. His team had not been looking for tetrapods, but Clack was hoping that somewhere in the mass of technical papers and field notebooks there would be signs of another tetrapod graveyard. She was in luck. Not only were fossils of tetrapods fairly common in some areas, the field notes said, but one member of the team had brought some fossils back to England. They were sitting in the basement of the Sedgwick Museum, where Friend worked.

When Clack tracked down the hunk of rock Friend's team had collected, she found that it contained three skulls. They were not *Ichthyostega*. A peculiar pair of horn-shaped projections at the back of the skulls identified them as *Acanthostega*, a 360-million-year-old early tetrapod that was known only from a few skull fragments Jarvik briefly described in 1952. This was a fantastic find, but Clack needed more complete skeletons, and fortuitously another team of scientists from Denmark were planning on further exploring the area from which the skulls had been exhumed. With the promising prospect of significant fossils still in the field Clack landed a spot on the expedition, and in 1987 she found even more remains of both *Acanthostega* and *Ichthyostega*.

To help her prepare these fossils Clack contacted vertebrate pale-ontologist Michael Coates, and they began to winnow the fossils out of the rock in 1989. Clack had already seen the complete skulls of *Acanthostega*, but as she and Coates peeled back the rock over the rest of the skeleton they began to find some very unusual features. The ra-dius and ulna—the two long bones between the wrist and upper arm bone—were of different lengths, with the bone on the "thumb side" of the arm, the radius, being much longer than its neighbor.

What the scientists found at the end of the arms was even more bizarre. It had long been assumed that the standard number of digits for tetrapods was five, and scientists expected that early tetrapods would either have five digits or would show signs that they had successively added digits until five fingers were formed. As Coates picked away at the hand of *Acanthostega* he did find five digits, but then he found more. When his work was done, eight fingers could clearly be seen on the hand of *Acanthostega*.

Acanthostega was not an aberration. The early tetrapod from Rus-sia, *Tulerpeton*, had six fingers, and a hindlimb of *Ichthyostega* Clack had brought back from Greenland had seven toes. Viewed together, the archaic hands and feet of these creatures seemed better suited for pro-pelling them through the water than crawling on land, and five fingers was clearly not the norm among the earliest tetrapods. Now paleon-tologists not only had to account for how fins were transformed into limbs, but how the polydactyl paddles of *Acanthostega* and *Ichthyostega* were transformed into a limb suitable for walking on land.

The pectoral ("shoulder") and pelvic ("hip") fins of the fleshy-finned fish *Eusthenopteron* formed the foundation of the change. Both pairs of fins were connected to the body by a single bone, the equivalent of the humerus in our arms and femur in our legs, and both sets were linked to each other through embryological development. Some of the earliest fish only had pectoral fins, but during their evolution a dupli-cation event occurred in which genes that regulated the development of the fish created a second pair of fins further down the body. This

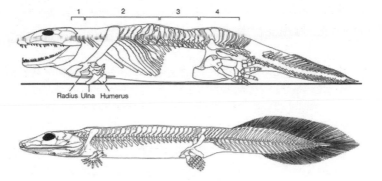

FIGURE 21 – The skeletons of *Ichthyostega* (top) and *Acanthostega* (bottom). The brackets above the *Ichthyostega* skeleton denote the different kinds of vertebrae: 1) cervical, 2) thoracic, 3) lumbar, 4) sacral.

ancient duplication event meant that these fins had a similar construction and were undergirded by the same genetic instructions.

The lower limbs of *Eusthenopteron* were more complicated. Bones that are homolgous to the radius and ulna in our limbs were present, but they were part of a branching out of bony elements that were surrounded by little rods of bone called lepidotrichia, which supported the margins of the fin. These bones were arranged along what embryologists called the metapterygial axis, a line of growth that runs from the parts of the bones nearest the body and branches out in the portions that are further away. This growth pattern is seen in living lungfish, and it was what gave the fins of fish like *Eusthenopteron* their distinctive shape.

The question was where this axis of growth could be drawn in the limbs of early tetrapods. While *Eusthenopteron* was not directly ancestral to early tetrapods, it was certainly a close relative and therefore useful as a model for the fish ancestors of tetrapods. As such, early tetrapod limbs would have grown along a metapterygial axis, too, but although the axis passed through the humerus (the only bone in the upper arm) its path afterward was unclear. This was an old question, one that anatomists had been debating since the nineteenth century, but the fossil record alone was insufficient to provide the answer.

The solution was to be found among living descendents of the earliest tetrapods. Despite some relatively minor tweaks, the overall form of the tetrapod limb has been widely conserved among living members of the group. Scientists hoped that by studying the limb development of living organisms they might also find clues as to how the limbs of early tetrapods evolved. This is precisely what paleontologist Neil Shubin and embryologist Pere Alberch did in 1986. What they found was very different from any growth scenario that had been proposed before.

At the time that Shubin and Alberch published their study it was well known that the limb bones of tetrapods started out as cartilage that condensed from the cells of the embryo. (Only later are the structures transformed into bone through a process called ossification.) This is how the parts of the limb closest to the body, the upper arm and leg bones, form. From there the nascent limb continues to grow and gives rise to the precursors of the lower limb bones (the radius and ulna of the arm, and tibia and fibula of the leg), which are separated from the upper limb bones by a joint.

Then the pattern of development shifts. The pairs of lower limb bones continue to grow while the radius and tibia split off a few bones of the hand and foot; the ulna and fibula begin to spout off bones arranged in an arc that sweeps across the bones of the wrist or ankle. The bones that are produced in this arc are the fingers and toes. Rather than running straight down the limb through a finger, this pattern revealed that the metapterygial axis became twisted, producing digits starting from the "pinky" side to the thumb/big toe side.[21] The ancient genetic and developmental link between the fore- and hindlimbs—the duplication that produced the pelvic fins in fish—meant that both hands and feet grew along these same pathways.

This pattern was supported by the observed action of Hox genes during limb development. Hox genes are like switches in an organism's DNA that regulate development and turn cascades of other genes on and off. Their most prominent function is organizing the head-to-tail organization of animal bodies, but there are also Hox genes that regulate the development of fins and limbs. Embryologists have traced the activity of some of these regulatory genes, such as Hoxd-11 and Hoxd-13, during fin and limb development in fish and tetrapods, and they show some surprising differences. In zebrafish, bony fish not very closely related to tetrapods or fleshy-finned fish like *Lepidosiren* the Hox genes are active along an axis that goes straight down the fin. In tetrapods, however, the same genes are active in the same straight-line pattern until they abruptly whip across the limb bud from the "pinky" side to the "thumb" side, the same metapterygial axis that Shubin and Alberch identified as giving rise to digits.

This twist made sense of why early tetrapods seemed to have extra fingers. During embryonic development there is part of the tetrapod limb bud on what will become the "pinky" side called the Zone of Polarizing Activity, or ZPA for short. This area produces a protein given the whimsical name Sonic Hedgehog, which travels to the edge of the limb bud, where it triggers the release of a hormone that stimulates growth. This hormone, in turn, tells the ZPA to make more Sonic Hedgehog and activates particular Hox genes that guide the development of digits. This developmental cycle produces the digits, from the "pinky" side of the hand or foot first, and the longer it goes on the more digits are produced. If it gets shut off early, fewer digits are formed, meaning that this developmental conveyor belt probably remained active longer in early tetrapods. A little tweak in development produced hands and feet instead of fins.[22]

The presence of hands and feet in *Acanthostega* did not mean that it was a terrestrial animal, however. Contrary to expectations, the overall constellation of features Clack and Coates found in the skeleton of *Acanthostega* were hallmarks of an aquatic animal. Its flattened skull was attached to its shoulders by way of a very short neck, so there was little space between the lower jaw and the front of the pectoral girdle. In fish, the bones that made up the shoulders were attached to the back of the skull, and while separated in *Acanthostega*, they were still closely associated. Its limbs were also short and its feet were broad, good for sculling through the water but not as well suited to walking. This was matched by the presence of bones in the tail that would have supported a broad paddle that would have helped propel *Acanthostega* through the water. As if that were not enough, *Acanthostega* still retained bony gill arches: it could have drawn oxygen directly from its watery habitat.

That *Acanthostega* probably had both gills and lungs is not altogether surprising. The most archaic forms of living bony fish, including lungfish, have both. Though such a dual system might seem unusual to us obligate air-breathers, it is essential to the survival of fish that inhabit waterways in which oxygen can often become depleted. Rather than representing a peculiar advancement toward tetrapods as nineteenth- century naturalists supposed, *Lepidosiren* and its allies may have more in common with more ancient kinds of fish.

Even though most bony fish do not have lungs they can breathe with, they do have modified lungs in the form of a swim bladder. Lungs did not evolve from swim bladders but rather the other way round: swim bladders in many familiar fish are a derived specialization that evolved long after the first tetrapods appeared. This suggests that many early fish, including the ancestors of tetrapods, probably had both gills and lungs. Some lineages retained lungs while others modified or lost theirs, but contrary to what we might expect, lungs are actually a very old vertebrate trait. Early tetrapods did not need to evolve lungs anew. They already had them thanks to their ancestry.

Like *Acanthostega*, *Ichthyostega* also had gill arches, lungs, a paddle-shaped tail, and an array of pressure sensors along its body called a "lateral line," but its skeleton also contained signs that it at least occasionally walked on land. *Ichthyostega* had a longer neck than *Acanthostega*, and its shoulders, situated farther back, bore huge shoulderblades that would have provided ample space for the muscles needed to deal with the stresses of terrestrial locomotion. *Ichthyostega* also had slightly differentiated types of vertebrae along the spine. In fish and earlier tetrapods the vertebrae are almost all identical since the sinuous side-to-side swimming patterns of these aquatic animals did not require much variation, but walking on land put new stresses on these bones. This resulted in modifications of the vertebral column; neck, upper back, lower back, hip, and tail vertebrae all differ in *Ichthyostega*, just as in our own skeletons.[23] Just how it might have moved on land is still unknown, especially since its overlapping ribs would have greatly constrained its movement from side to side, but the details of its anatomy suggest that *Ichthyostega* was a much more terrestrial animal than its cousin.

Together both *Acanthostega* and *Ichthyostega* showed that many traits thought to mark life on land had evolved in the water first. Tetrapods had not evolved as a result of fish wriggling over baking soil as Lull and Romer had suggested, and geologists realized that the rust-colored rocks in which the first were found did not represent the arid world that Barrell had envisioned. *Acanthostega* and *Ichthyostega* lived in shallow, swampy environments that were not under threat of evaporating completely. Fins had evolved into limbs and fingers in these aquatic habitats; rather than being a consequence of moving on land they were a preexisting condition for it.

The tetrapod story had suddenly been altered, and the discovery of the complete skeletons of these tetrapods prompted paleontologists to scour museum collections to see if there were any more hiding in forgotten cabinets. These searches turned up the fragmentary remains of early creatures like *Elginerpeton*, *Ventastega*, and other members of an early radiation of tetrapods. But old problems still remained. There was still nothing between *Eusthenopteron* and the earliest known tetrapods to document how fins were transformed into limbs. Anatomy and embryology confirmed that it must have happened, but there was still a gap in the fossil record.

As before, long-neglected fossil specimens would play an important role in fleshing out early tetrapod history. Despite being named in 1942, the fleshy-finned fish *Panderichthys* had for decades only been known from fragments. It did not seem to hold much relevance to the questions

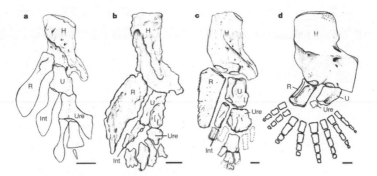

FIGURE 22 – A comparison of the forelimbs of *Eusthenopteron*, *Panderichthys*, *Tiktaalik*, and *Acanthostega*. Abbreviations: *H*, humerus; *Int*, intermedium; *R*, radius; *U*, ulna; *Ure*, ulnare. Scale bar, 1 cm.

surrounding early tetrapods, and even when a partial skeleton of the fish was dug out of a quarry in Latvia in 1972 it languished in storage, undescribed, for years. As paleontologists with an eye for early tetrapods dove back into old collections in search of new material, however, *Pandericthys* stood out.

Though classified as a fish, *Panderichthys* very closely resembled early tetrapods. From nose to tail it had a flattened body well suited to life in the shallows, and its fins were reminiscent of those in the earliest known tetrapods. It was not a terrestrial animal, but its limbs would have allowed it to push its upper body off the bottom to gulp air from the surface.

But *Panderichthys* was soon overshadowed by another creature. During the 1990s Neil Shubin, the paleontologist who had teamed up with Pere Alberch to explain the development of the tetrapod limb, was studying the Devonian-age rocks of Pennsylvania with his student Ted Daeschler. Together they helped to describe the fragmentary remains of a new tetrapod found in those rocks they called *Hynerpeton*, but it was nowhere near as complete as the *Acanthostega* fossils Clack had dug out of Greenland. Looking for early tetrapods in Pennsylvania was difficult business. Most of the deposits of the right age and type were covered by soil or sat under suburban developments, rendering them inaccessible. The few available deposits had been exposed as a result of department of transportation crews blasting through hills to clear the way for roads, often destroying fossils in the process. Even though the Pennsylvania sites had potential, getting anything out of them was a massive headache. Shubin and Daeschler resolved to look elsewhere.

The first step in fossil hunting is pinpointing the most likely place to

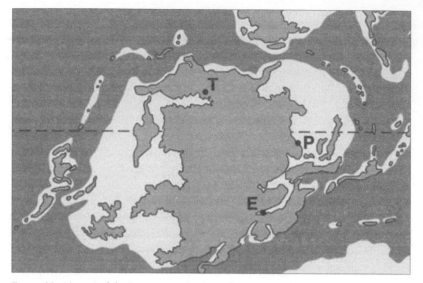

FIGURE 23 – A map of the Euramerican landmass during the late Devonian. The dashed line indicates the equator. Sites of elpistostegalian fish discoveries: *T*, *Tiktaalik*; *E*, *Elpistostege*; *P*, *Panderichthys*.

find your quarry, so Shubin and Daeschler began flipping through textbooks hoping to find Devonian deposits of the right age and type to hold early tetrapods. During this process a dispute broke out about some bit of geological trivia that only another textbook could resolve, but when the scientists opened one to continue their argument they were struck by something they had somehow missed. There was a swath of Devonian rocks of just the kind they were looking for in the upper latitudes of Canada that had not yet been picked over by other paleontologists.

The 1999 expedition of the Nunavut region of Canada led by Shubin and Daeschler did not yield the fossils they had been hoping for. The next year they decided to move to a site near the southwestern tip of Ellesmere Island, but their search only turned up well-known varieties of Devonian fish. They kept searching, each expedition a gamble, but they failed to hit the payoff they had been expecting. It was decided that the 2004 expedition would be their last try.

They were in luck. Much to the delight of the paleontologists, encased in the approximately 375-million-year-old rock were the remains of a flat-headed, long-snouted fish unlike any the scientists had seen before. When they started to remove it from the encasing rock back in the United States they found that they had recovered the fairly complete remains of a creature that further documented the evolution of limbs.

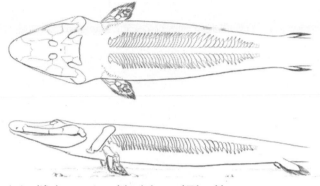

FIGURE 24 – A simplified restoration of the skeleton of *Tiktaalik*.

Dubbed *Tiktaalik roseae* in 2006, this creature was a tetrapodomorph that was even more closely related to *Acanthostega* than *Panderichthys* was. *Tiktaalik* had a similar low profile well suited to life in the shallows, but unlike *Panderichthys* it also had a neck. The head and shoulder girdles were separated, allowing *Tiktaalik* to move its head without having to move its body (a useful trait for snapping up fish, which its teeth suggested it did), but the parts of *Tiktaalik* that garnered the most attention were its arms.

The well-preserved forelimbs of *Tiktaalik* were built on the familiar humerus–radius and ulna–fin radials–lepidotrichia pattern, but it possessed rudimentary joints between its fin radials and other arm bones that would have allowed it to flex its forelimbs. It was still very much an aquatic creature, but these intricate traits suggested that it could better support its body weight in the shallow-water environment in which it lived. As the authors of the first description of *Tiktaalik* concluded, "New discoveries of transitional fossils such as *Tiktaalik* make the distinction between fish and the earliest tetrapods increasingly difficult to draw."

Tiktaalik also provided another reference point with which to compare other early tetrapods. When more complete skull and shoulder material of *Ventastega* was described by Per Ahlberg and colleagues in 2008 the scientists were able to determine that this creature was part of a diversification of tetrapod types. *Ventastega* was probably not a direct descendant of *Tiktaalik* or an ancestor of *Acanthostega*, but part of a radiation of forms that lived in near-shore environments near the close of the Devonian. The fin-to-limb transition was better documented than ever before, but the fossils could not be slotted into a straight lineup of ancestors and descendants. This not only confirmed that evolution has a branching pattern, but it also suggested that there are many more

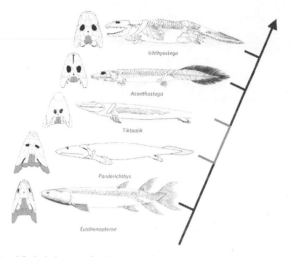

FIGURE 25 – A simplified phylogeny of early tetrapods and their close relatives. The diagram does not indicate a direct line of descent, but successive branching points that illustrate the changes early tetrapods went through during the evolutionary transition.

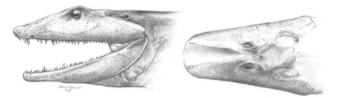

FIGURE 26 – The restored head of *Ventastega* as seen from the side (left) and above (right).

species of early tetrapod waiting to be discovered.[24]

    Ultimately many of these early tetrapod forms succumbed to extinction, but they also formed the basis of a radiation of the first amphibians. One lineage among these varied forms, in turn, gave rise to the first amniotes—lizardlike vertebrates that could reproduce by laying shelled eggs further away from the water's edge. As disparate as these creatures were, though, they all were modifications of the ancestral tetrapod body plan, themselves modified from fleshy-finned fish. The traits that allowed these radiations of terrestrial vertebrates, from lungs to fins supported by a bony architecture, were in place for millions of years before the first tetrapod crawled on land. This co-option would become a recurring theme in the evolution of vertebrates, including one particular lineage of tetrapod descendants, birds.

Footprints and Feathers on the Sands of Time

"*Tempora mutantur, et mutandum est avibus* (The times have changed, and the birds must change with them)"
—EDWARD HITCHCOCK, *Ichnology of New England*, 1858

"There is no evidence that *Compsognathus* possessed feathers; but, if it did, it would be hard indeed to say whether it should be called a reptilian bird or an avian reptile."
—T. H. HUXLEY, *American Addresses*, 1877

Humans have been finding the traces of extinct creatures for thousands of years. Unaware of their true identity, a variety of cultures have inter preted fossil footprints, shells, and bones as the remnants of gods, heroes, saints, and monsters. The cyclops, griffins, and numerous other beings of myth and legend were not just figments of human imagination but monsters restored from the remains of creatures dead for millions of years. It was no different among the Native Americans of North America. The Tuscaroras, Iroquois, Onondagas, and many other tribes had legends inspired by fossils, including the Lenape of the Delaware Valley.[25]

At the time Europeans arrived in North America the Lenape occupied the land from northern Delaware to the Hudson Valley of New York, and in the blood-red sandstone of this range they saw three-toed, clawed footprints.[26] According to one tale, passed down by Richard Calmet Adams, some of these were said to be the footprints of the primeval progenitor of all the great monsters of the land and sea. It was a living horror, the destroyer of all it could dig its claws into, but perished when it was trapped in a mountain pass and obliterated by lightning.

Europeans that settled in the Connecticut Valley noticed the tracks, too. While plowing his father's field in South Hadley, Massachusetts, around 1802 a young man named Pliny Moody turned up slabs of rock indented with weird footprints. At least one of these curiosities was appropriately put to use as a doorstep, and visitors to the Moody farm sarcastically remarked that the Pliny's family must have raised some hearty chickens if they left footprints in solid stone. The physician Elihu Dwight, who later bought the slab, had a different interpretation. To him the tracks were made by Noah's raven when the biblical Deluge subsided.

Such tracks were hardly unique. The abundant three-toed footprints were often called "turkey tracks" (although many indicated turkeys bigger than a full-grown human), and a cache of the impressions were discovered by laborers quarrying flagging stones near Greenfield, Massachusetts, in 1835. These were brought to the attention of the local physician James Deane, who knew they were not made by antediluvian poultry or biblical birds. Just what had created them, though, Deane could not say, and so he contacted Amherst geology professor Edward Hitchcock and Yale academic Benjamin Silliman for their opinions.

Hitchcock was initially skeptical of Deane's claims. Some mundane geological phenomenon could have produced tracklike marks, the professor cautioned, but Deane was adamant that the footprints were genuine. Deane sent Hitchcock a cast of one of the footprints to support his case, and despite his doubts Hitchcock was intrigued. Hitchcock soon set out to have a look at the Greenfield tracks for himself and found that Deane was right. The impressions were the footsteps of ancient creatures that had trod the Connecticut Valley long before humans had settled there.

Hitchcock became enthralled by the tracks. He collected and purchased as many as he could. He fancied himself a scientific pioneer. Although Deane was also researching the tracks, Hitchcock was the first to publish on them in an 1836 issue of the *American Journal of Science*.[27] There was a variety of footprint types, each given a unique binomial name to indicate a different species, but the three-toed ones were some of the most remarkable. They ranged from giant tracks over seventeen inches long to tiny impressions less than an inch from front to back. A few large slabs even showed the strides of the animals, and the only reasonable conclusion was that they had been made by birds that flocked along the ancient shoreline. "Four out of five, I presume, would draw this conclusion at

FIGURE 27 – One of the many track-bearing sandstone slabs that enthralled Edward Hitchcock.

once," Hitchcock noted, and he thought that the tridactyl footprints were made by extinct equivalents of storks and herons that strode along the banks of an ancient lake or river.

Hitchcock was deeply inspired by the varied assemblage of birds that had once lived in the Connecticut Valley, and he attempted to do justice to his fascination in the anonymously published poem "The Sandstone Bird." In the geologist's verse, science is placed in the guise of a sorceress who conjures up the most majestic of the primeval birds:

> Bird of sandstone era, wake!
> From thy deep dark prison break.
> Spread thy wings upon our air,
> Show thy huge strong talons here:
> Let them print the muddy shore
> As they did in days of yore.
> Pre-adamic bird, whose sway
> Ruled creation in thy day,
> Come obedient to my word,
> Stand before Creation's Lord.

So restored, Hitchcock's fictional bird could only lament the dismal state of the modern world. The earth was cold and the impressive giants it knew so well were all gone. Even the trees were so Lilliputian that the dinosaur "*Iguanodon* could scarce here find a meal!" The haughty bird could not stand the sight of what had become of its home.

> . . . all proclaims the world well nigh worn out,
> Her vital warmth departing and her tribes,
> Organic, all degenerate, puny soon,
> In nature's icy grave to sink forever.
> Sure 'tis a place for punishment designed,
> And not the beauteous happy spot I loved.
> These creatures here seem discontented, sad:
> They hate each other and they hate the world,
> I can not, will not, live in such a spot.
> I freeze, I starve, I die: with joy I sink,
> To my sweet slumbers with the noble dead.

The sullen bird was then swallowed up by the earth, leaving the geologist with no evidence to prove what he had seen.[28] Hitchcock was in a similar bind. No skeleton had been found to reveal the true form of his birds. Storks and herons provided fair analogs, but even the largest of the living wading birds was puny compared to the birds that made the largest fossil tracks. Without skeletons, Hitchcock could only guess what they looked like.

At the same time that Hitchcock was researching the Connecticut Valley tracks, Richard Owen was examining a strange chunk of bone from New Zealand. It was said to have belonged to an enormous eagle, but Owen took it to be part of the femur of a gargantuan, ostrichlike bird he called *Dinornis* (commonly known as the moa). From the osteological scrap he reconstructed an entire skeleton, and it was later proven to be correct when more remains of the flightless birds were found. Owen had raised a giant bird from the dead, and it provided the perfect proxy for the sandstone birds.

For Hitchcock, though, there were more than just scientific lessons to be learned from the tracks. What he saw in the fossil record spoke of God's benevolence, and he expounded upon this belief as a Congregationalist pastor and professor of natural theology at Amherst. (Part of his inspiration for collecting so many tracks was to build a testament to God's glorious works in nature.) He was astonished by the vast array of stupendous creatures that crawled, swam, flew, and dashed over the surface of the earth in time immemorial. Though facts from the geological strata were shaking the foundations of a literal interpretation of Genesis, Hitchcock attempted to bridge the gap between geology and theology as the *Bridgewater Treatises* had in England. In his *Ichnology of New England* Hitchcock concluded:

> And how marvellous the changes which this Valley has undergone in its inhabitants! Nor was it a change without reason. We are apt to speak of these ancient races as monstrous, so unlike existing organisms as to belong to another and quite different system of life. But they were only wise and benevolent adaptations to the changing condition of our globe. One common type runs through all the present and the past systems of life, modified only to meet exigencies, and identifying the same infinitely wise and benevolent Being as the Author of all. And what an interesting evidence of his providential care of the creatures he has made, do these modifications of structure and function present! Did the same unvarying forms of organization meet us in every variety of climate and condition, we might well doubt whether the Author of Nature was also a Providential Father. But his parental care shines forth illustriously in these anomalous forms of sandstone days, and awakens the delightful confidence that in like manner he will consult and provide for the wants of individuals.

If God provided for birds that could neither sow nor reap their own food surely He would have also cared for the enormous avians of old (and even more so the human "lords of creation"). Hitchcock believed that only God could have so perfectly fitted organisms to their surroundings, but this view of nature crumbled as naturalists increasingly tried to understand nature on its own terms and not as a moral lesson. Charles Darwin's 1859 treatise slammed the door shut on the concept of natural theology as science, which Hitchcock subscribed to, but this new perspective on life's history raised new questions.

Birds were so different from other vertebrates that they appeared to be perched on their own lonely branch in the tree of life. How could they have evolved? Hitchcock's tracks hinted that true birds had been present nearly as long as reptiles and amphibians, and the discovery of a fossil feather in 1860 from Solnhofen, Germany, did nothing to change this quandry. Found in the Jurassic-aged limestone of a quarry mined for stone to make lithographic plates, the delicate fossil was acquired by the German paleontologist Christian Erich Hermann von Meyer. In 1861 he named it *Archaeopteryx lithographica*, the "ancient feather from the lithographic limestone."

Not long after von Meyer described the feather, another nearby limestone quarry produced an enigmatic skeleton. The jumbled creature had a long bony tail but was surrounded by feather impressions; it was as much a reptile as it was a bird. Rather than going straight to a museum, however, the specimen was given to the local physician Karl Häberlein in exchange for medical services.

Rumors of the specimen began to circulate among naturalists, but Häberlein would not part with it easily. He stipulated that the fossil would only be sold along with the rest of his fossil collection, raising the cost beyond the reach of many prospective buyers. Richard Owen and George Robert Waterhouse, certain that *Archaeopteryx* would bring prestige to the British Museum, were able to convince the trustees of the institution to forward £700 for the fossil (or what the museum would normally have spent on new fossil acquisitions over the course of two years). By November 1862 the fossil was in London.

Some German naturalists were upset that the slab had been expatriated to England, but the august University of Munich professor Johann Andreas Wagner had opposed efforts to acquire *Archaeopteryx* for his college. He was sure it was not all it seemed. Although Häberlein tried to restrict access to the specimen amid rumors it was a fake, a verbal report and sketch of the fossil reached Wagner, who argued that rather than a bird, it was a kind of reptile he called *Griphosaurus*, or "riddle reptile."

Wagner's fears over evolution had spurred his impulsive description. *Archaeopteryx* sounded like just the type of transitional form that would throw support to Darwin and Wallace's evolutionary theories, and Wagner's warnings about the fossils were among the last of his publications before his death.

Owen's description of the fossil was read before the Royal Society in 1863. He appraised it as the "by-fossil-remains-oldest-known feathered Vertebrate." More than that, the fossil was most certainly a bird despite its reptilian characteristics, and Owen upheld von Meyer's original name *Archaeopteryx*.[29] This diagnosis allowed Owen to make a particular prediction. The head of *Archaeopteryx* was missing, but Owen reasoned that "by the law of correlation we infer that the mouth was devoid of lips, and was a beak-like instrument fitted for preening the plumage of *Archaeopteryx*."

While some naturalists felt that Owen's description was rather crude, the news of the fossil was welcome among evolutionists. In an 1863 letter to Darwin the fossil mammal expert Hugh Falconer beamed,

> Had the Solenhofen quarries been commissioned—by august command—to turn out a strange being à la Darwin—it could not have executed the behest more handsomely—than in the *Archaeopteryx*.

This news made Darwin eager to hear more about the "wondrous bird," yet he ultimately did little to present *Archaeopteryx* as a confirmation of his evolutionary ideas. In the fourth edition of *On the Origin of Species* published in 1866, Darwin primarily used *Archaeopteryx* and Hitchcock's tracks—by now thought to have been made by dinosaurs—to illustrate that the fossil record still had secrets to divulge. "Hardly any recent discovery," Darwin wrote of *Archaeopteryx*, "shows more forcibly than this how little we as yet know of the former inhabitants of the world." Even as it hinted at a connection, *Archaeopteryx* was too weak to unequivocally bridge the gap between reptiles and birds by itself. The necessary evidence would be supplied by the anatomist Thomas Henry Huxley.

Huxley began his scientific career in 1846 by studying marine invertebrates while serving as an assistant surgeon aboard the HMS *Rattlesnake*. His work was well received by other naturalists, and when he returned to England in 1850 he was set to establish himself among the scientific elite. Like the man who would become his rival, Richard Owen, Huxley was most concerned with the underpinnings of anatomical form, but where Owen cloaked his work in pious rhetoric, Huxley's distaste for religious interference in science may have attracted him to

FIGURE 28 – Thomas Henry Huxley, photographed around 1870.

Darwin's theory of evolution in the first place. While Huxley disagreed with Darwin on some key points, natural selection was the best mechanism yet proposed for evolutionary change. For natural selection to make sense, however, the absence of graded transitions in the fossil record had to be accounted for, which Huxley explained through the concept of "persistent types."

Throughout the fossil record there seemed to be little evolutionary change; crocodiles looked like crocodiles no matter what strata they came from. Instead of being evidence *against* evolution, however, Huxley proposed that the persistent forms were echoes of evolutionary changes that had occurred in a time so distant that it was not recorded in the rock. If most of evolution happened during "non-geologic time," then the inability of naturalists to explain the origin of major groups of animals with fossil evidence became a moot point. The caprices of geology kept them out of science's reach.

This concept was a double-edged sword. It removed the problem of missing transitional forms but it made it nearly impossible to determine evolutionary relationships through fossil evidence. But Huxley was not concerned with drawing out ancestors. Instead, he was after the common denominators of animal form, and birds and reptiles provided a key example of how the same plan could be modified to different ends. During his 1863 lectures on vertebrate anatomy at the Royal College of Surgeons, Huxley asserted that birds were "so essentially similar to Reptiles in all the most essential features of their organization, that these animals may be said to be merely an extremely modified and aberrant Reptilian type." Reptiles, too, shared similarities with birds, and to reinforce these connections Huxley placed both birds and reptiles into an encompassing group called the "Sauropsida" (thus labeling birds "reptile-faced").

Huxley reiterated this point in his 1867 survey of birds. Reptiles and birds were modifications of the same "groundplan," with living reptiles being closer to the hypothetical framework from which each had been adapted. If one were to compare a turtle with a dove, this association might seem laughable, but it was not among the lowly lizards and snakes that the best evidence for the connection between reptiles and birds was to be found. The solution of a fossil puzzle provided a better set of candidates.

While traveling through England that same year Huxley met geologist John Phillips, who invited Huxley to visit the museum at Oxford with him. As the naturalists strolled through the geology collection Huxley noticed something strange about the bones of the dinosaur *Megalosaurus* on display. A portion of its shoulder blade was actually part of the hip. Once this scrap was put in its proper place other fragments caught Huxley's eye. When the two scientists finished reorganizing the bits of bone, they found they had restored a predator with small forelimbs and a birdlike pelvis.[30] This new shape for *Megalosaurus* pointed to a deeper relationship between reptiles and birds that Huxley had inadvertently been amassing evidence for since his Royal College lectures. Dinosaurs were much more birdlike than any living reptiles, and inspired by the branching evolutionary trees in the 1866 book *Generelle Morphologie* by the German embryologist Ernst Haeckel, Huxley started to think of how birds actually could have evolved from reptiles. In January 1868 Huxley outlined a preliminary line of descent in a letter to Haeckel.

> In scientific work the main thing just now about which I am engaged is a revision of the *Dinosauria*—with an eye to the *Descendenz Theorie*! The road from Reptiles to Birds is by way of *Dinosauria* to the *Ratitae*—the Bird "Phylum" was Struthious, and wings grew out of rudimentary fore limbs.

Huxley would unveil this evolutionary trajectory to his peers later that same year. Even though there was no direct evidence for the transition he was proposing, the forms that had already been found suggested that the connection between birds and reptiles was real. *Archaeopteryx*, for instance, was clearly a bird with reptilian characteristics. He conceded that it was "more remote from the boundary-line between birds and reptiles than some living *Ratitae* [flightless birds such as ostriches and emus] are," and therefore not a direct ancestor of modern birds, but it still illustrated the point that birds could have evolved from reptiles. While the complete evolutionary series had yet to be found, the anatomical resemblances between flightless birds and fossil creatures like *Megalosaurus* suggested that

FIGURE 29 – Two visions of *Megalosaurus*. While originally envisioned as an immense crocodile-like beast, as shown by the restoration on the left, by the latter part of the nineteenth century naturalists had greatly revised the appearance of the dinosaur, as shown by the restoration to the right. Unfortunately, since so little is known of *Megalosaurus*, we can only base our ideas of what it looked like on related theropod dinosaurs.

the first birds had been derived from something resembling a dinosaur. This was made possible by a major change in the way paleontologists understood dinosaurs.

The first dinosaurs known to science, *Megalosaurus* and *Iguanodon*, were initially thought to have looked like enormous crocodiles and lizards. So little was known of them that they were easily cast as larger versions of known reptiles, but when Richard Owen grouped them within the Dinosauria in 1842 he gave them an anatomical overhaul. Dinosaurs, as he envisioned them, were warm-blooded creatures that carried their limbs directly beneath their bodies. They were the "highest" of the reptiles, much more impressive than their degenerate reptilian kin that inhabited the modern world, but the fragmentary nature of their remains left most of their anatomy uncertain.[31] The discovery of a more complete dinosaur revealed that dinosaurs looked strikingly different from what Owen envisioned.

Found in the sandy marl of New Jersey in 1858, and later described by William Parker Foulke and Joseph Leidy, *Hadrosaurus* was a Cretaceous herbivore related to *Iguanodon*. Unlike Owen's reconstruction of *Iguanodon*, however, its skeleton suggested that it walked upright at least some of the time. A predatory dinosaur from nearby deposits called *Laelaps* by its discoverer E. D. Cope (later renamed *Dryptosaurus* by his rival O. C. Marsh) also shattered Owen's dinosaurian archetype. This New World relative of *Megalosaurus* walked on two legs, and the fact that the forelimbs of the animal were much shorter than the hind limbs caused Cope to envision an active, hot-blooded dinosaur that relied on its powerful hind limbs to kill:

This relation [between the hind limbs and forelimbs], conjoined with the massive tail, points to a semi-erect position like that of the Kangaroos, while the lightness and strength of the great femur are altogether appropriate to great powers of leaping. . . . If he were warm-blooded, as Prof. Owen supposes the Dinosauria to have been, he undoubtedly had more expression than his modern reptilian prototypes possess. He no doubt had the usual activity and vivacity that distinguishes the warm-blooded from the cold-blooded vertebrates.

We can, then, with some basis of probability imagine our monster carrying his eighteen feet of length on a leap, at least thirty feet through the air, with hind feet ready to strike his prey with fatal grasp, and his enormous weight to press it to the earth. Crocodiles and Gavials must have found their bony plates and ivory no safe defence, while the Hadrosaurus himself, if not too thick skinned, as in the Rhinoceros and its allies, furnished him with food, till some Dinosaurian jackalls dragged the refuse off to their swampy dens.

If *Hadrosaurus* and *Dryptosaurus* walked on two legs it was reasonable that *Iguanodon* and *Megalosaurus* could have done the same. Three-toed tracks from the same deposits that yielded *Iguanodon* were in accord with the idea that it was bipedal at least some of the time, and Huxley's own revision of *Megalosaurus* at Oxford suggested that it also stalked about on two legs. Yet these animals presented a substantial problem for Huxley's evolutionary program. They were enormous animals, far too large to be good models for the forerunners of birds.

The chicken-sized dinosaur *Compsognathus* was a far better candidate for the sort of creatures from which birds evolved. Discovered in

FIGURE 30 – *Compsognathus*, as restored in Huxley's American addresses.

1861 from the same quarries that yielded *Archaeopteryx*, it was more birdlike than any of its gargantuan relatives, especially in details of its hind limbs and ankles. This similarity had been recognized by the German anatomist Carl Gegenbaur in 1864, and even the anti-evolutionist Wagner drew attention to it in his description of the animal; but where Wagner disavowed that the similarities were evidence for evolution, *Compsognathus* was Huxley's prime evidence that birds had sprung from reptiles. Speculating upon what it might have looked like in life, Huxley wrote:

It is impossible to look at the conformation of this strange reptile and to doubt that it hopped or walked, in an erect or semi-erect position, after the manner of a bird, to which its long neck, slight head, and small anterior limbs must have given it an extraordinary resemblance.

With this new vision of dinosaurs in place, Huxley continued to accumulate evidence that birds had been derived from the dinosaur body plan. The small dinosaur *Hypsilophodon*, while less birdlike than *Compsognathus*, was significant as it provided Huxley with the first good look at a complete dinosaurian pelvis. The process that normally extended forward in reptiles, the pubis, was rotated backward to meet the ischium, as in birds. Huxley thought it reasonable that all dinosaurs had this arrangement, and he also appealed to embryology to imply that at certain states developing chicks exhibited dinosaurlike traits.

If the whole hind quarters, from the ilium to the toes, of a half-hatched chicken could be suddenly enlarged, ossified, and fossilized as they are, they would furnish us with the last step of the transition between Birds and Reptiles; for there would be nothing in their characters to prevent us from referring them to the *Dinosauria*.

The English paleontologist Harry Seeley criticized this interpretation. The congruence between the hind limbs of birds and dinosaurs could be attributed to a shared mode of life, Seeley argued, and not a family relationship. In Seeley's view, walking bipedally on land had caused the legs of both dinosaurs and birds to take similar form, and thus the resemblance was only skin-deep. This was particularly significant as Seeley had specialized in studying another group of reptiles that he thought were closer to birds.

The first pterosaur known to science was discovered in 1784 in a German limestone quarry. With a tooth-studded snout, lizardlike hind limbs, and a ludicrously long fourth finger on each hand, the creature was unlike any that had been seen before. The man who described it, Italian naturalist Cosmo Alessandro Collini, thought it was a swimmer, since it had come from marine deposits. Others disagreed and proposed that it was closely related to bats, but in 1809 Georges Cuvier recognized it as a unique kind of extinct flying reptile. He dubbed it *Pterodactylus*, or "wing finger."

Not everyone was in agreement with Cuvier. In 1830, the German researcher Johannes Wagler reconstructed the animal as something of a cross between a swan and a penguin, which sculled about the surface of the water with a paddle supported by the elongated finger. Another specimen discovered in 1828 by fossil hunter Mary Anning was investigated by William Buckland. The creature was clearly a reptile, but Buckland was perplexed by its features, and he thought that, like Milton's "Fiend" in *Paradise Lost*, the pterosaur could have swum, sunk, waded, crept, or flown through a strange ancient world. By the 1840s, however, there was little doubt that Cuvier had been correct, and some naturalists were very impressed by resemblances between the skeletons of the flying fiends and birds. As Richard Owen stated in an 1874 monograph of Mesozoic fossil reptiles:

> Every bone in the Bird was antecedently present in the framework of the Pterodactyle; the resemblance of that portion directly subservient to flight is closer in the naked one to that in the feathered flier than it is to the forelimb of the terrestrial or aquatic reptile.

Just like Owen, Seeley saw no way to "evolve an ostrich out of an *Iguanodon*," but Huxley turned the argument from convergence against his opponents. The traits supposedly shared between birds and pterosaurs had to do with flight, and given that both lineages had become adapted to flying, common traits in their skeletons were to be expected. The diagnostic traits in the hips, legs, and feet of dinosaurs, on the other hand, were found in all birds, not just ground-dwelling ones. This meant that these characters marked a true family relationship and not just a shared way of life.

To formalize this new image of dinosaurs Huxley placed them in new taxonomic groups to underline their avian characteristics. The dinosaurs and *Compsognathus* (which Huxley considered to be the closest relative to dinosaurs but not one itself) were put together under the name Ornithoscelida, making them the "bird-legged" members of the "reptile-faced" Sauropsida. Yet, despite all the work he had done on the topic, Huxley could not rule out any of the dinosaurs then known to be bird ancestors. Some represented the form the real ancestors may have taken, but that was all.

Huxley explained this argument in an 1870 presidential address before the Royal Society. In searching for evolutionary lineages, Huxley warned, "it is always probable that one may not hit upon the exact line of filiation, and, in dealing with fossils, may mistake uncles and nephews for fathers and sons." To prevent this sort of confusion he drew a distinction between intercalary types, or representations of the *form* of ancestors and descendants, and linear types, which were the actual ancestors and descendants.

At the present moment we have, in the *Ornithoscelida* the intercalary type, which proves that transition ["from the type of the lizard to that of the ostrich"] to be something more than a possibility; but it is very doubtful whether any of the genera of *Ornithoscelida* with which we are at present acquainted are the actual linear types by which the transition from the lizard to the bird was effected. These, very probably, are still hidden from us in the older formations.

After 1870 Huxley's paleontological work slowed. He was in over his head delivering lectures, writing papers, and engaging in the politics of science—so much so that he burned himself out. His wife, Nettie, sent him on a vacation to Egypt in 1872 with the hope that he would recover from the stress, and when Huxley returned he started on a new tack. He turned his attention to the minutiae of anatomy under the microscope, largely setting aside the old bones that had previously transfixed him.

But Huxley did not abandon the evolution of birds entirely. In 1876 he set out on a lecture tour of the United States, and one of his first stops was Yale's Peabody Museum run by American paleontologist O. C. Marsh. Though little new information about the origin of birds had been found since the time of Huxley's 1870 address, Marsh had recently found the remains of toothed birds in the Cretaceous-age chalk of Kansas. One of the birds, *Hesperornis*, had tiny nubs for wings and looked like a loon with a tooth-studded beak; the other, *Ichthyornis*, would have looked more like a toothed gull in life.[32]

Marsh's odontornithes ("toothed birds") strengthened the link between reptiles and birds, but they were geologically too young to indicate from what group birds had evolved. Along with *Archaeopteryx* and *Compsognathus*, and the early Jurassic dinosaurs which made the Connecticut Valley tracks, they could not be placed on a straight evolutionary line but instead signaled what Huxley believed was an earlier transition:

It is, in fact, quite possible that all these more or less avi-form reptiles of the Mesozoic epoch are not terms in the series of progression from birds to reptiles at all, but simply the more or less modified descendants of Palaeozoic forms through which that transition was actually effected.

FIGURE 31 – The reconstructed skeleton of *Hesperornis*. As a bird with teeth, it further confirmed the connection between birds and reptiles that Huxley highlighted.

> We are not in a position to say that the known *Ornithoscelida* are in-
> termediate in the order of their appearance on the earth between reptiles
> and birds. All that can be said is that if independent evidence of the actual
> occurrence of evolution is producible, then these intercalary forms remove
> every difficulty in the way of understanding what the actual steps of the
> process, in the case of birds, may have been.

Despite the numerous strands of evidence Huxley had tied together, the question of avian origins was far from settled, especially as his hypothetical trajectory of avian evolution came under attack. There was a growing consensus that flightless birds had evolved from flying ancestors. If this was the case, the ratites could not be used as examples of what early birds had been like. Indeed, even though the "intercalary types" identified by Huxley were important to considerations of avian evolution, there was no consensus as to how they related to each other.

Naturalists tried to make sense of the tangle of data in different ways. German paleontologist Carl Vogt proposed that flightless birds evolved from dinosaurs while flying birds evolved from pterosaurss. His colleague Robert Wiedersheim endorsed a modified version of this idea. Georg Baur, by contrast, thought that the backward-pointing hips of dinosaurs like *Hypsilophodon* and *Iguanodon* pinned them as ancestral to birds. One of Huxley's pupils, E. Ray Lankester, expressed his belief that birds had evolved from aquatic dinosaurs and had wings derived from flippers.[33]

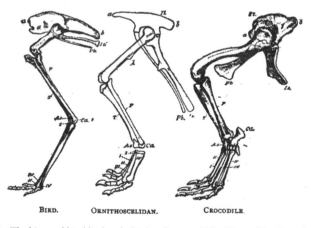

BIRD.   ORNITHOSCELIDAN.   CROCODILE.

FIGURE 32 – The hips and hind limbs of a bird, a dinosaur ("Ornithoscelidan"), and a crocodile, as presented by Huxley in his American addresses. Huxley used this diagram to stress the bird-like nature of the hind limbs of dinosaurs.

A second, more exquisitely preserved *Archaeopteryx*, discovered in 1877, fueled these continuing debates. Found in a quarry in Eichstätt, not far from Solnhofen, it is arguably the most beautiful fossil ever discovered. Whereas the "London specimen" was scrambled, the new specimen was fully articulated, its head thrown back and arms spread wide to display a splash of feathers. The fact that it had a head greatly increased its significance. Although the first specimen appeared to have been decapitated in 1865, John Evans thought that he had discovered a portion of its toothed mouth on the same slab as the rest of the skeleton. Some said that the jaws belonged to a fish, but Evans did not think it unreasonable that a bird with so many reptilian characteristics would also have teeth. The new specimen confirmed Evans's hypothesis and refuted that of Owen. *Archaeopteryx*, like *Hesperornis* and *Ichthyornis*, had tooth-studded jaws. This confirmation fed into the ongoing debates of the creature's affinities, but regardless of what group it was assigned to it was such an enigmatic fossil that it could not be ignored. In time it would be agreed that it was the very first bird, a creature that documented one point in one of life's major transformations.

There was more to the debate than anatomy and family trees, however. The origin of birds was directly tied to questions about the origins of flight, and an early attempt to tackle this problem was undertaken in 1879 by paleontologist Samuel Williston. Taking a dinosaurian ancestry for birds as a starting point, Williston proposed:

It is not difficult to understand how the fore legs of a dinosaur might have been changed to wings. During the great extent of time in the Triassic, in which we have scanty records, there may have been a gradual lengthening of the outer fingers and greater development of the scales, thus aiding the animal in running. The further change to feathers would have been easy. The wings must first have been used in running, next in leaping and descending from heights, and, finally, in soaring.

A similar idea was later developed by the eccentric Hungarian aristocrat, spy, and paleontologist Baron Franz Nopcsa von Felső-Szilvás. He proposed that while pterosaurs evolved from quadrupedal ancestors that lived in the trees and took to the air by leaping, birds had evolved from terrestrial predecessors that jumped and "oared along in the air" with the help of feathered arms.

Yet the "ground-up" origin for flight hypothesized by Williston and Nopcsa failed to gain a firm foothold, and other researchers continued to mull over how flight could have originated. A particularly ingenious solution to the problem was proposed by the American ornithologist William Beebe. Despite the fact that Beebe thought *Archaeopteryx* more of a "flutterer" than a true flyer, he believed that it might represent an early stage of flight, and he used it as a starting point to predict what its ancestors and descendants might look like.[34]

Beebe introduced his colleagues to his hypothetical transitional series in 1915. It had all started in the trees. As Beebe had observed in the New World tropics, iguanas sometimes leapt out of trees when frightened, and when they did so they flattened themselves out to slow their descent. In such a scenario longer scales would increase their surface area to further slow their falls, Beebe reasoned, especially if these scales were situated along the arms. But the back end of the animal would have had to have been held up, too, otherwise it would be akin to a reptilian Darius Green, who, like the subject of John Townsend Trowbridge's poem, would fall "to the ground with a thump! Flutt'rin' an' flound'rin', all'n a lump!"

The key to how these hypothetical creatures stayed aloft was found in living birds. A recently hatched dove Beebe examined had rudimentary feather quills attached to its upper leg, and *Archaeopteryx* appeared to have long feathers on its legs, too. Thus, Beebe surmised, the ancestors of birds had leg wings that helped balance them out while

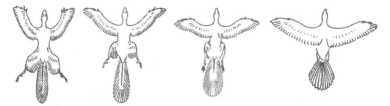

FIGURE 34 – William Beebe's hypothesis of the evolution of birds. According to Beebe's scenario, bird ancestors would have started by parachuting in a "Tetrapteryx stage," and over time the feathers of the front wings would have become enlarged, allowing for powered flight.

parachuting and had gone through a "Tetrapteryx stage." As the scales turned into real feathers and the animals became capable of gliding the forewings became more prominent, and the feathers of the tail became larger to support to back of the body. By combining fossil evidence with studies of living animals, Beebe was able to make a functional prediction of how birds had evolved.

Beebe's hypothesis was only one competing among many, though, and no clear consensus could be drawn. Naturalists were unable to confirm whether flight had evolved from the "trees down" or "ground up." Without knowledge of the ancestral form any hypothesis could be constructed from the scraps of evidence.

Older and more primitive than even the most ancient dinosaurs, the pseudosuchians seemed like good candidates for the ancestors of pterosaurs, dinosaurs, and birds. As proposed by paleontologist Robert Broom, whereas dinosaurs had peculiar specializations that would bar them from being birds ancestors, the psuedo-suchians such as *Euparkeria* were still "generalized" creatures from which both groups could have easily derived. This would make any resemblances between birds and dinosaurs matters of convergence and not real signals of ancestry.

FIGURE 35 – The skull of *Euparkeria*.

The Danish artist Gerhard Heilmann most forcefully articulated this hypothesis in his 1926 book *The Origin of Birds*. Some dinosaurs closely resembled birds, particularly the coelurosaurs such as the predatory *Gorgosaurus* and ostrichlike *Struthiomimus*, but they lacked one characteristic that barred them from a close relationship to birds: clavicles. According to Dollo's Law, which was formulated by Belgian paleontologist Louis Dollo, evolution could not be reversed, and therefore dinosaurs could not be bird ancestors as it would require that they regrow clavicles after they had already lost them.[35] This left the pseudosuchians the most appropriate stock from which to derive birds.

Heilmann's work was a classic, and the pseudosuchian origin for birds became the favorite hypothesis during the following four decades. It was so widely accepted that even when clavicles were described among the remains of the small, predatory dinosaur *Segisaurus* in 1936 no one seemed to notice. (The first specimen of *Oviraptor* described in 1923 also had clavicles, but they were misidentified at the time.) The problem of bird origins had been solved; all that was needed were fossils to confirm the transition.

With the big question of bird ancestry seemingly resolved, work on the subject slowed during the middle of the twentieth century. Occasional alternate interpretations of *Archaeopteryx* continued to pop up, some closely linking the bird to dinosaurs, but the pseudosuchian hypothesis remained the favored one. Still, the resemblance between birds and predatory dinosaurs was undeniable. The immense sauropod dinosaurs were often considered to be drab, tail-dragging animals that spent much of their time in swamps, but the small predatory dinosaurs were another matter. Writing in the middle of the twentieth century, paleontologist Edwin Colbert thought the theropod *Ornitholestes* was an "agile" catcher of lizards and insects, and its compatriot *Ornithomimus* had "very long, slender hind limbs and very birdlike feet, which indicate that it must have been a rapid runner, much as are the modern ostriches."

It would take the rediscovery of a dinosaur first found in 1931 for paleontologists to begin to fully realize the significance of the theropod dinosaurs to bird evolution. During the summer of 1964 paleontologists John Ostrom and Grant E. Meyer from Yale's Peabody Museum were searching for fossils near the town of Bridger, Montana, when they discovered the numerous fragments of an unusual dinosaur. The famous fossil hunter Barnum Brown had found the remains of the same kind of dinosaur, which he informally called "Daptosaurus," decades earlier, but since he never fully described it few paleontologists knew anything about it. Based upon the more complete remains they had found, however, Ostrom and Meyer knew that Brown had overlooked a dinosaur unlike any other then known.

They called the new predator *Deinonychus* ("terrible claw"), so named because of the wicked, sickle-shaped weapon it carried on its

FIGURE 36 – A modern restoration of *Deinonychus*.

second toe. The arrangement of the bones showed that *Deinonychus* held this claw off the ground, and the tail of the animal was stiffened by ossified rods that would have acted as a dynamic counterbalance. This was not a slow, stupid predator but an agile predator, and the presence of multiple individuals from the same site associated with bones of the herbivorous dinosaur *Tenontosaurus* suggested that *Deinonychus* might have been a pack hunter, something practically unheard of in dinosaurs.[36] Of *Deinonychus*, Ostrom wrote:

> *Deinonychus* must have been anything but "reptilian" in its behavior, responses and way of life. It must have been a fleet-footed, highly predaceous, extremely agile and very active animal, sensitive to many stimuli and quick in its responses. These in turn indicate an unusual level of activity for a reptile and suggest an unusually high metabolic rate.

Deinonychus stood in sharp contrast to the traditional image of dinosaurs. Even though nineteenth-century naturalists like Owen, Cope, Huxley, and Seeley thought that dinosaurs were warm-blooded animals, the consensus since that time had shifted to envision dinosaurs as larger versions of living lizards and crocodiles. Like their living counterparts they would have required a warm environment in order to be active, but the details of their physiology were unknown. What was supposed about their biology had been inferred from living reptiles in studies like that carried out on alligators by Edwin Colbert, Charles Bogert, and Raymond Cowles in 1946.

In order to approximate dinosaurian physiology, the trio of scientists carried out the unenviable task of sticking thermometers in the cloacae of American alligators. Several specimens, ranging from one to seven feet in length, were placed in the sun or shade and had their temperature taken every ten minutes. (Larger animals would have been better, but as the researchers explained, "the difficulties of making temperature experiments [on fully grown alligators] would be great and can be best left to the imagination.") What the scientists found was that the larger alligators warmed up and cooled down slower than the smaller ones. It took about a minute and a half for the small animals to warm up one degree Celsius, while it took the largest animals five times as long to do the same. This was regulated by their internal volume. As the size of a body or object increases, its internal volume increases exponentially. An ostrich egg, for instance, is only about two and a half times as large as a chicken's egg, yet it contains about twenty times as much fluid and tissue inside. (If you wanted to make a hard-boiled ostrich egg it would take much, much longer for the heat to cook it than it would for a chicken's egg.) Likewise, the larger alligators had more internal volume and so took a greater amount of time to heat up or cool down. Extrapolating these differences up to the size estimates for dinosaurs, the authors wrote that it would take a ten-ton dinosaur around three and a half days of basking out in the sun to raise its body temperature one degree Celsius!

But as the researchers found out the hard way with two of their test animals, prolonged exposure to the hot sun could be deadly. It was absurd to think that dinosaurs had to sunbathe for so long to become active. (They revised their figures in later publications, writing that a large dinosaur would have to spend most of one day heating up, but this was still an unreasonable amount of time to spend sunbathing.) It was more likely that the large size of many dinosaurs shielded them from fast heat fluctuations, and that they benefited from a high, stable body temperature that would have allowed them to be active much of the time.

This only made sense for the largest dinosaurs. At only one meter tall *Deinonychus* was too small to maintain a near-constant high body temperature, yet it was adapted for a very active life. Was it possible that some dinosaurs maintained a high body temperature internally? Ostrom and his student Bob Bakker thought so, and French paleontologist Armand de Ricqlès came to a similar conclusion almost simultaneously through his work on the microstructure of dinosaur bone. This launched a lively, and sometimes acrimonious, debate about the lives of dinosaurs.

After simmering for several years, the debate over "hot-blooded dinosaurs" came to a boil during a 1978 symposium hosted by the American Association for the Advancement of Science. While no clear consensus could be reached, it was apparent that the phrases "warm-blooded" and "cold-blooded" were easily misused. A better understanding of the physiology of many different organisms revealed a wide diversity of metabolic strategies that were not easily categorized. An animal that controls its body temperature internally, maintains that high temperature regardless of external temperature, and has a high metabolic rate while at rest is called "endothermic." Animals traditionally called "cold-blooded," on the other hand, do not have constant, internally regulated body temperatures. Their metabolic rates can be high or low depending on external factors, giving them the label "ecotherms," and they can be just as active as endothermic animals under the right conditions.[37]

The question that remained was whether dinosaurs were endotherms or ectotherms, but without living subjects to observe it was difficult to know for sure. As paleontologist Peter Dodson opined, it was perhaps best to consider "dinosaurs as dinosaurs." But what if dinosaurs did have living descendants, after all? The discovery of *Deinonychus* and the debate over dinosaur physiology reinvigorated interest in the idea that birds had evolved from dinosaurs, and if this was correct then the physiology of birds would be a model for understanding the lives of dinosaurs.

A key piece of new evidence in this reinvestigation came from a mislabeled specimen in a museum. In 1855, five years before the first *Archaeopteryx* feather was found, Hermann von Meyer acquired what appeared to be a pterosaur skeleton from the German limestone quarries. When Ostrom saw it over a century later, however, he knew it was no pterosaur. It was a specimen of *Archaeopteryx* that had been misidentified, and it was strikingly similar to *Deinonychus*. After carefully studying the "new" specimen, Ostrom came to the same conclusion English zoologist Percy Lowe had arrived at in 1936 (albeit by a different route). "The osteology of *Archaeopteryx*, in virtually every detail, is indistinguishable from that of contemporaneous and succeeding coelurosaurian dinosaurs," Ostrom wrote, confirming that the first bird was a theropod dinosaur.[38]

The revival of the avian dinosaur hypothesis was not immediately well received. The pseudosuchian hypothesis still held strong, even as the pseudosuchia (now sometimes called thecodontia) was recognized as a taxonomic wastebasket that did not constitute a natural evolutionary group. Slowly, however, many paleontologists came around to the view that birds might be the direct descendants of dinosaurs, even as the fossils that would confirm the transition remained elusive.

If Ostrom was right that coelurosaurs gave rise to birds, then it was likely that there were other feathered theropods waiting to be discovered.[39] The likelihood of finding feathered dinosaurs, however, was slim. Even under the best of circumstances fossil preservation is a capricious thing. Fully articulated skeletons are rare, and rarer still are fossils that preserve any indication of body covering or soft tissues.

It was for just this reason that a snapshot circulated at the 1996 Society of Vertebrate Paleontology meeting held at the American Museum of Natural History caught paleontologists off guard (John Ostrom among them). It showed a little theropod dinosaur not unlike *Compsognathus* with its head thrown back and tail pointed straight up, and along its back was a strip of fuzzy feathers. Although no scientific study had yet been undertaken (the fossil had only come to the attention of Canadian paleontologist Phil Currie and paleo-artist Michael Skrepnick two weeks earlier), the specimen confirmed the connection between dinosaurs and birds that had been proposed on bones alone. The new dinosaur was dubbed *Sinosauropteryx*, and it had come from Cretaceous deposits in China that exhibited a quality of preservation that exceeded that of the Solnhofen limestone.

Sinosauropteryx was only the first feathered dinosaur to be announced. A panoply of feathered fossils started to turn up in the Jurassic and Cretaceous strata of China, each just as magnificent as the one before. There were early birds that still retained clawed hands (*Confuciusornis*) and teeth (*Sapeornis, Jibeinia*), while non-flying coelurosaurs such as *Caudipteryx, Sinornithosaurus, Jinfengopteryx, Dilong,* and *Beipiaosaurus* wore an array of body coverings from wispy fuzz to full flight feathers. The fossil feathers of the strange, stubby-armed dinosaur *Shuvuuia* even preserved the biochemical signature of beta-keratin, a protein present in the feathers of living birds, and quill knobs on the forearm of *Velociraptor* reported in 2007 confirmed that the famous predator was covered in feathers, too.

As new discoveries continued to accumulate it became apparent that almost every group of coelurosaurs had feathered representatives, from the weird secondarily herbivorous forms such as *Beipiaosaurus* to *Dilong*, an early relative of *Tyrannosaurus*. It is even possible that,

FIGURE 37 - A *Velociraptor* attempts to catch the early bird *Confuciusornis*. Both were feathered dinosaurs.

during its early life, the most famous of the flesh-tearing dinosaurs may have been covered in a coat of dino-fuzz.

The coelurosaurs were among the most diverse groups of dinosaurs. The famous dinosaurs *Velociraptor* and *Tyrannosaurus* belonged to this group, as did the long-necked, pot-bellied giant herbivore *Therizinosaurus* and birds. What is remarkable is that, with the exception of the ornithomimosaurs, every branch on the coelurosaur family tree contains at least one feathered dinosaur, and it is expected that fossils of even more feathered coelurosaurs will be discovered as investigations continue. This suggests that, instead of evolving independently in each group, feathers were a shared trait for coelurosaurs that was inherited from their common ancestor.[40] Most, if not all, coelurosaurs probably had some kind of feathery covering for at least part of their lives.[41]

A mix of fossil and molecular evidence hints at how feathers could have evolved. Birds are the living descendants of the coelurosaurs, and crocodylians are the closest living relatives to dinosaurs as a whole, so features shared between birds and crocodylians might have been present in the last common ancestor of both lineages (and therefore also present in dinosaurs). Both birds and alligators, for example, share the regulatory proteins sonic hedgehog (abbreviated Shh, and named for the video game character) and bone morphogenetic protein 2 (BMP2–), both of which underlie the formation of both the scales of alligators and the feathers of birds. Hence it is likely that, during the evolution of dinosaurs, these proteins were co-opted from their roles in forming the tough hides of dinosaurs into the creation of feathers.

The diversity of feather types among coelurosaurs suggests how feathers were modified once they had begun to evolve. As seen in *Sinosauropteryx*, the earliest feathers were simply tubes that grew from the skin. Once these structures evolved there would have been enough variation for them to split and become branched, something that has been observed in the downy covering of baby chickens, with each feather providing greater coverage on the animal. From there, the branching filaments could be organized along a central vane, like what is seen in *Caudipteryx* and *Sinornithosaurus*. After this point, little barbs branched off from each filament along the shaft, locking them together and stiffening the feather. This was the kind of feather needed for flight, and it is what is seen in most modern birds. That these structures are feathers and not just degraded collagen or some other quirk of fossilization is beyond reasonable doubt.

The majority of dinosaur fossils are just bones and teeth, and even fossilized skin impressions only preserve patterns, not colors.

But scientists have recently discovered that there is a way to detect some colors in the fossil record. While studying an exceptionally preserved fossil, squid paleontologist Jakob Vinther saw that its ink sac was packed with the same kind of microscopic spheres that give the ink of living squid their color. These bodies are called melanosomes, and once Vinther realized that they could be preserved in the fossil record he began to wonder what other prehistoric remains might contain them.

One of the first tests was on the forty-seven-million-year-old feather of an extinct bird from Messel, Germany (home of "Ida" and not far from the final resting place of *Archaeopteryx*). Since the feather seemed to show light and dark bands, it was a good test case to see whether the bodies were truly pigment-carrying melanosomes (in which case they would only be found in the dark bands) or were just bacterial remnants scattered all over the feather. The results were better than could have been expected. In 2009 the researchers behind the study announced that not only did the feather most certainly contain melanosomes in the dark bands, but their arrangement corresponded to a pattern seen in living birds that gives feathers a glossy sheen. This was better than just an isolated discovery. It presented paleontologists with a new technique and two teams, working independently of each other, turned to the fossilized feathers of dinosaurs to see if they, too, contained remnants of color.

The first team, lead by Fucheng Zhang, published their results in the journal *Nature* on January 27, 2010. They had turned their attention to two of the first feathered dinosaurs to be found, *Sinosauropteryx* and *Sinornithosaurus*. Feather samples from both contained two different types of melanosomes; those that created dark shades (eumelanosomes) and those associated with reddish hues (phaeomelanosomes). This allowed the scientists to speculate that *Sinosauropteryx* had a garish red-and-white-striped tail, which might have been used to signal to other members of its species.

Vinther and his team published their own findings in *Science* the following week. Building on the previous research on the fossil bird feather, they attempted to present a specimen of the recently discovered dinosaur *Anchiornis* in Technicolor. After determining the pattern of melanosome distribution throughout the feathers, they compared the arrangements to what is seen in living birds to restore the long-lost pigments. As it turned out most of the feathers of *Anchiornis* were black, but they were set off by white accents on its wings and a plume of rufous feathers on top of its head. Even though the study did not look

FIGURE 38 – Restorations (not to scale) of *Anchiornis* and *Microraptor*, based upon exceptional specimens that also preserved feathers. The discovery of such fossils has overwhelmingly confirmed that birds evolved from dinosaurs.

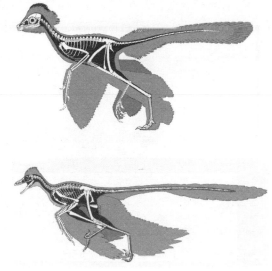

for chemical traces of color in the fossil that would have marked the presence of other shades, for the first time the researchers were able to produce an image of an entire living dinosaur.

The question of just what a feather is has become more complicated, however. Very early on in dinosaur evolution there was a split in the dinosaur family tree that resulted in the evolution of the ornithischians (containing an array of herbivorous dinosaurs such as the ankylosaurs, hadrosaurs, and ceratopsians) and the saurischians (comprised of the predatory theropods and the forebears of the gigantic, long-necked sauropods).[42] The presence of feathers in coelurosaurs alone suggested that fuzzy body coverings had evolved only once among dinosaurs within the saurischian side of the split, but at the beginning of the twenty-first century scientists found similar structures among ornithischian dinosaurs. In 2002 Gerlad Mayr and colleagues announced that they had discovered a specimen of the ceratopsian *Psittacosaurus* with long, bristlelike structures growing out of its tail and it was joined in 2009 by *Tianyulong*, another bristle-covered ornithsichian described by a team of researchers led by Zheng Xiao-Ting.

These animals were about as far removed from bird ancestry as it was possible to be while still remaining a dinosaur, yet they were covered in structures similar to the proto-feathers of *Sinosauropteryx*. Either the filamentous body covering evolved twice in two different groups of dinosaurs, or, even more spectacularly, was a common di-

FIGURE 39 – A *Styracosaurus*, covered in bristles, scavenges the body of a dead tyrannosaur. The discovery that ornithischian dinosaurs like *Tianyulong* and *Psittacosaurus* had bristlelike structures growing out of their skin suggests that it is possible that many other ornithischian dinosaurs did, as well.

nosaur trait later lost in some groups. Regardless of how many times "dino fuzz" evolved, however, these structures were only adapted into true feathers among the coelurosaurs, but how flight evolved is another evolutionary mystery.

John Ostrom presented one hypothetical scenario in 1979. Inspired by his work on *Deinonychus* and *Archaeopteryx*, he proposed that the ancestors of the first bird were small coelurosaurs covered in rudimentary feathers. With their grasping hands, these tiny predators would have been adept hunters of flying insects, and their simple feathers would have provided an unexpected advantage. The feathers along their arms would have helped trap insects, and so longer feathers would have been selected for over time. Eventually these "proto-wings" would have allowed the dinosaurs a little bit of extra lift while jumping after their prey, and this shift in selection would precipitate the origin of the first flying birds.

Ostrom's "insect-net hypothesis" never truly took off, as it was marred by functional problems surrounding how feathers might be used as a net, but it did reignite an old debate about whether flight evolved from the "trees down" or the "ground up." According to the advocates of the arboreal hypothesis, small feathered dinosaurs climbed up into trees and launched themselves into the air to glide a short distance, and eventually they would be adapted to beat their wings to truly fly. The four-winged dinosaur *Microraptor*, a relative of *Deinonychus*, has most recently been taken to throw support to this idea, as it may have launched itself out of trees to glide, if not truly fly, through the forest.

Other paleontologists have preferred one version or another of the cursorial hypothesis. In this view, feathered dinosaurs ran along the ground, perhaps hopping into the air after insects or other prey, until by some mechanism they developed the ability to actually fly. In fact, feathered arms may have even made some dinosaurs better runners. A key piece of evidence for this hypothesis comes from chukar partridges. These birds are capable of flight, but if they need to escape into a nearby tree or over a natural obstacle they often run rather than fly, flapping their wings as they do so. As discovered by scientist Kenneth Dial this technique gives the birds better traction while running, so much so that they can run right up vertical inclines. As hypothesized by Dial, feathered dinosaurs could have gained a functional advantage by flapping their arms while running (be it after prey or to avoid becoming prey), and this behavior could then be co-opted to allow them to start flying.

As recognized by most working paleontologists today, however, the old arboreal versus cursorial dichotomy is no longer helpful. Much like Williston, Nopsca, and Beebe, we can create numerous plausible scenarios but, without knowing which feathered dinosaurs were the root stock from which birds evolved, any origin-of-flight hypothesis must be regarded as provisional right from the start. Even as the numerous feathered fossils have confirmed that birds evolved from dinosaurs, they have also made the relationships between those fossils and birds much more complex. At one time it seemed that *Velociraptor* and its relatives were the closest relatives of early birds, but a little-known group of recently discovered forms may be even closer.

Described in 2002, the small feathered dinosaur *Scansoriopteryx* was one of the most bizarre coelurosaurs ever found. With large eyes, a short snout, and a very long third finger, this sparrow-sized dinosaur was unlike many of its coelurosaur cousins. Its description was followed in 2008 by the announcement of a close relative named *Epidexipteryx*, a pigeon-sized dinosaur, covered in fuzz, that also sported two pairs of ribbonlike feathers on its shortened rump and a mouth full of forward-oriented teeth. Given that they may be older than the earliest birds, they could represent the kind of dinosaur birds evolved from, in which case the *Velociraptor* and its relatives would be further removed from the origin of birds than had been previously supposed.

For over a century *Archaeopteryx* was the key to understanding

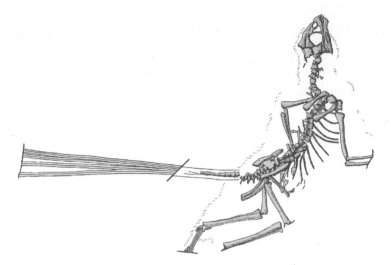

FIGURE 40 — A drawing of the skeleton of *Epidexipteryx*, denoting the "halo" of feathers around the skeleton and the pairs of elongated feathers coming out of its tail. It may be one of the closest relatives of early birds.

bird origins, as it was the oldest bird ever discovered, but as more feathered dinosaurs have been found the connection between *Archaeopteryx* and other fossil birds has become looser. As the delineation between non-avian dinosaur and bird has become increasingly blurred it has become difficult to tell what side *Archaeopteryx* falls on. As research continues, it may turn out that *Archaeopteryx* was, like *Microraptor*, a feathered dinosaur and not a true bird.

The unstable relationships of some of the feathered dinosaurs was exemplified by the redescription of *Anchiornis huxleyi* in 2009. The fossil, named in honor of T. H. Huxley's work on bird origins, had been announced the year before as the closest dinosaurian relative of birds and, at thirty million years older than *Archaeopteryx*, was especially significant. When a better-preserved specimen was found, however, the scientists realized that their initial hypothesis was wrong. *Anchiornis* was actually a troodontid, or a member of a group of coelurosaurs closely related to the famous "raptors," yet it was very similar in form to *Archaeopteryx*.

Even if *Archaeopteryx* is dethroned from the vaunted position of "earliest known bird" that Richard Owen bestowed upon it, the fact remains that birds evolved from dinosaurs, and much more than fossilized feathers supports this hypothesis. During the 1920s the explorer

Roy Chapman Andrews led a string of expeditions for the American Museum of Natural History into Mongolia's Gobi Desert to search for the evolutionary center of origin for all mammals (including humans). No evidence of a mammalian Eden was found, but the excursions did return with the ghostly white bones of the Cretaceous dinosaurs *Velociraptor*, *Protoceratops*, and *Oviraptor*, the latter of which was especially fascinating because it was found in the act of robbing a *Protoceratops* nest.

But in 1994 it was announced that the wrong dinosaur was inside the supposed *Protoceratops* eggs. Instead of an embryonic horned dinosaur, there was the miniscule skeleton of a developing theropod much like *Oviraptor*. The specimen the Andrews expedition had found was probably caring for its own eggs, not robbing the eggs of others. The discovery of several skeletons of a crested *Oviraptor* relative named *Citipati*, which were found sitting atop nests of the same kind of eggs, supported this hypothesis. Their arms encompassed the sides of the nest in a position only seen in birds, and the close relationship between *Citipati* and the feathered *Caudipteryx* opened the possibility that these dinosaurs, too, were covered in feathers that they used to regulate the temperature of their nests. This discovery of fossilized behavior dovetailed beautifully with the numerous feathered coelurosaurs, and the description of the tiny troodontid *Mei long* in 2004

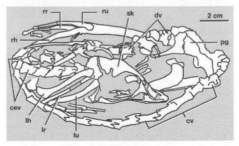

also surprised paleontologists. Like the skeletons of *Citipati* on their nests, several of these dinosaurs were suddenly killed and buried while sleeping, perfectly preserved in the position in which they died. They were curled up just like slumbering birds.

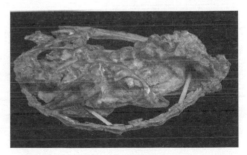

FIGURE 41 – The skeleton of the troodontid dinosaur *Mei long*, with a line drawing identifying the visible bones on the right. Abbreviations: *cev*, cervical vertebrae; *cv*, caudal vertebrae; *dv*, dorsal vertebrae; *lh*, left humerus; *lr*, left radius; *lu*, left ulna; *pg*, pelvic girdle; *rh*, right humerus; *rr*, right radius; *ru*, right ulna; *sk*, skull.

The unique breathing system seen in modern birds also appeared long before their ancestors first took to the air. As you relax reading this book you go through a breathing cycle of inhaling and exhaling. When you inhale, air enters your lungs (where oxygen is absorbed), and when you exhale the carbon dioxide-rich, oxygen-depleted air is forced out. Unlike you, however, birds lack a diaphragm and cannot inflate or deflate their lungs. Instead birds have a "one way" breathing system in which fresh air moves through their respiratory system both when the bird inhales and exhales. This is made possible by a series of anterior and posterior air sacs that can expand and contract. This is a more efficient way of getting oxygen from the air, but these air sacs also have a structural benefit. They arise from the lungs and invade the surrounding bones, thus making birds lighter. This infiltration into the bone leaves telltale hollows and indentations on bones, which have been seen in dinosaurs for over one hundred and fifty years.

It might not come as a surprise that coelurosaurs have evidence of air sacs on their bones, but other saurischian dinosaurs shared the same feature, too. This makes sense given the evolutionary history of these dinosaurs. There is no sign that air sacs were present in the ornithischian dinosaurs, but the evidence for air sacs in saurischian dinosaurs goes all the way back to one of the earliest presently known. Called *Eoraptor*, this small bipedal dinosaur was not unlike *Compsognathus*,

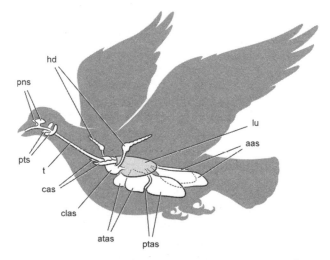

FIGURE 42 – A diagram of the air sacs inside a bird. Abbreviations: *atas*, anterior thoracic air sac; *cas*, cervical air sac; *clas*, clavicular air sac; *hd*, humeral diverticulum of the clavicular air sac; *lu*, lung; *pns*, paranasal sinus; *ptas*, posterior thoracic air sac; *pts*, paratympanic sinus; *t*, trachea.

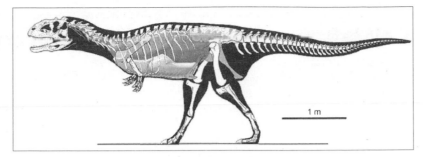

FIGURE 43 – A reconstruction of the skeleton of *Majungasaurus*, showing the placement of air sacs within the body inferred from pockets in its bones. Although *Majungasaurus* was not closely related to birds, the presence of these structures in its skeleton shows that these features were widespread among saurischian dinosaurs.

and it may be a fair approximation of what some of the earliest saurischian dinosaurs were like. Its bones were marked by indentations that indicate that it had at least some rudimentary air sacs, and later predatory dinosaurs from the coelurosaurs to the knobbly-headed abelisaur *Majungasaurus* and the *Allosaurus*-relative *Aerosteon* had even better developed air sacs.

The other great saurischian dinosaur group, the sauropods, also had bones infiltrated by air sacs. If you tried to design an animal like a 100-foot-long sauropod with thick, heavy bones in its neck, it would have been unable to lift its head. Much like a bridge, their skeletons reflect the selective pressures for strength and lightness, and air sacs allowed them to achieve this. They probably inherited this trait from their last common ancestor with the theropod dinosaurs.

While not exactly like those seen in living birds, the air sacs in many of these saurischian dinosaurs may have also provided them physiological benefits. Air sacs may have initially been selected because they lightened the skeleton, but if they provided dinosaurs with more efficient breathing (allowing them to be more active, for example) there would have been additional benefits for natural selection to act upon. Research into this area is still new, but it is clear that rudimentary air sacs appeared in dinosaurs seventy-five million years before *Archaeopteryx*, long preceding the first birds.

Some dinosaurs were even plagued by parasites that now infest the mouths of living birds. From healed wounds on skulls paleontologists have known for years that large predatory dinosaurs bit each other on the face during combat. Tyrannosaurs, especially, showed scars from such conflicts, but many *Tyrannosaurus* jaws often had holes in the lower jaw not apparently caused by the teeth of a rival. When paleontologists Ewan Wolff, Steven Salisbury, Jack Horner, and David Varricchio took another look at the jaws of tyrannosaurs that had these holes they did not find any sign of infection, inflammation, or healing that would be expected if the dinosaurs had been bitten. Bone, after all, is living tissue, and would slowly remodel itself in the wake of an injury. Instead, the holes were smooth, as if the bone was being slowly eaten away.

It seemed more likely that the holes were the result of some kind of pathology, and the researchers found that the sores were consistent with damage done by a single-celled protozoan called *Trichomonas gallinae* that infests modern birds. When inside living birds this microscopic creature causes ulcers to form in the upper digestive tract and mouth of the host, virtually identical to the damage seen in the *Tyrannosaurus* jaws. The species of protozoan that afflicted *Tyrannosaurus* might have been only a close relative of the living kind, but this was the first evidence of an avian disease afflicting dinosaurs.

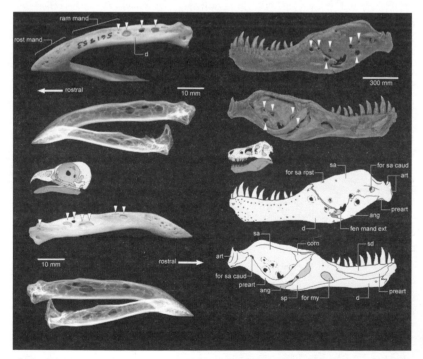

FIGURE 44 – The lower jaw of a hawk compared with the lower jaw of a *Tyrannosaurus rex*, both showing lesions in the bone caused by the microorganism *Trichomonas gallinae*.

Traits we think of as clearly identifying birds—feathers, air sacs, behavior, and even peculiar parasites—were present in a wide variety of dinosaurs first. Distinguishing the first true birds from their feathered dinosaur relations has become increasingly difficult. If we define birds as warm-blooded, feathered, bipedal animals that lay eggs, then many coelurosaurs are birds, so we have to take another approach.

Living birds, from kiwis to chickadees, fall within the group Aves, which also includes extinct birds like *Confuciusornis*, *Jeholornis*, *Zhongornis*, *Lonipteryx*, *Hesperornis*, and *Archaeopteryx*. Overall, Aves is the taxonomic equivalent of what are often informally referred to as "birds," but the earliest birds share many features with their closest relatives among the non-avian dinosaurs. What the closest dinosaurian relatives of birds may be, though, is presently under debate. Deinonychosaurs, the group containing both the dromaeosaurs (i.e. *Deinonychus*, *Microraptor*) and troodontids (*Mei*, *Anchiornis*), has often had pride of place as the dinosaurs nearest to birds and the group from which birds evolved. The identification of *Archaeopteryx* as a feathered dromaeosaur certainly reinforces this view, but the research describing the dinosaurs *Scansoriopteryx* and *Epidexipteryx* has placed them even closer to birds than the dromaeosaurs.

If the new analyses are supported by further evidence, *Scansoriopteryx* and *Epidexipteryx* together would make up a group called the Scansoriopterygidae and be the closest relatives to Aves. Thus, Aves plus the Scansoriopterygidae would form a group called the Avialae, with the deinonychosaurs being the next closest relatives to both groups. This placement does not reveal direct ancestors and descendants, but rather represents the group of dinosaurs from which birds arose and what they might have looked like. It is extremely unlikely that a direct line of descent from birdlike dinosaur to the first dinosaur-like bird will be found.

In his 1871 critique of evolution by natural selection, *On the Genesis of Species*, George Jackson Mivart considered the wings of birds a damning example of how Darwin's theory failed. To him a bird's wing was an atrophied organ, degenerate in the number of digits and bones in each finger. "Now, if the wing arose from a terrestrial or subaerial organ, this abortion of the bones could hardly have been serviceable— hardly have preserved individuals in the struggle for life." In other words, how could organisms have survived with half-formed wings?

What we know now about evolution has undermined Mivart's contention. The limbs of birds are only the modified limbs of dinosaurs; all the bones in the wing of a bird were present in the terrible, grasping

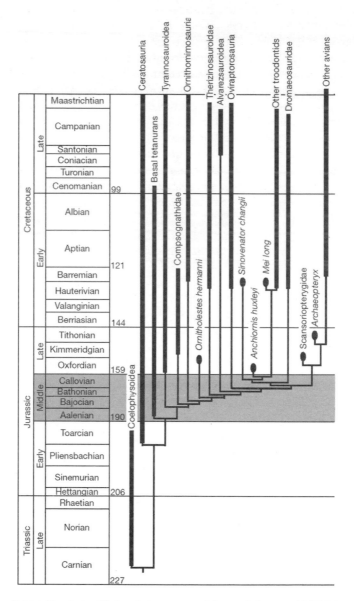

FIGURE 45 – A simplified evolutionary tree of theropod dinosaurs highlighting the relationships of coelurosaurs and birds.

hands of *Deinonyonus* and the delicate manus of *Epidexipteryx*. There is scarcely anything about a pigeon perched on a statue or a chicken you eat for dinner that did not first appear in dinosaurs, long before *Confuciusornis* flew in great flocks over what is now China. The majority of their relatives sunk into extinction sixty-five million years ago, but they are perhaps the most successful dinosaurs ever to have evolved. If you want to see living dinosaurs, you don't have to trek to a steaming jungle or isolated plateau. All you have to do is put up a bird feeder and look out the window.

But dinosaurs and birds were not the only terrestrial vertebrates evolving during the Mesozoic. The first mammals evolved alongside early dinosaurs, but they remained small creatures that lived in the corners of the world's ecosystems. The worst mass extinction ever to strike the planet had almost entirely wiped out their ancestors, making them only the remnants of a family that once flourished, but, 150 million years later, a stroke of bad luck for the dinosaurs would prove to be an unexpected boon.

The Meek Inherit the Earth

"The relatively late time at which mammals took over the world's supremacy from the reptilian dynasties would lead one to think that the stock from which they sprang must have been developed at a comparatively late date in reptilian history. This, however, is exactly the reverse of the true situation."
—ALFRED SHERWOOD ROMER, *Vertebrate Paleontology*, 1933

Scotland did not have much to offer nineteen-year-old Andrew Geddes Bain. Both his parents had died when he was a child, and even though he was educated his job prospects were few. So when his uncle, Lieutenant Colonel William Geddes, left for South Africa in 1816, young Andrew tagged along.

Once he arrived in the Cape colony Bain worked as a saddler, an explorer, an ivory trader, a soldier, and a road builder, but in 1837 he read a book that inspired him to look a little bit closer at the rugged landscape around him. It was Charles Lyell's influential *Principles of Geology*, and just like young Charles Darwin, Bain was smitten with Lyell's work. It allowed him to see the traces of lost worlds right beneath his feet, and his job as a road builder gave him the chance to see in the field what Lyell had described in print.

As he searched the dusty, shrub-flecked landscape of South Africa's Karoo desert for fossils one day in 1838, Bain discovered a skull unlike any he had seen before. The petrified creature had a short, turtlelike head complete with a beak, but instead of being toothless it possessed two large tusks that jutted down from the upper jaw. It was a fantastic find, and its relatively good state of preservation made it all the more spectacular. Most fossils from the Karoo were crushed, distorted, extremely fragile, and encased in nearly impenetrable sandstone.

Bain determined that the skull had come from a hitherto unknown type of "bidental" animal, but he did not have the background in anatomy to fully comprehend what he had found. To find out more, Bain sent the skull to the Geological Society in England in 1844.

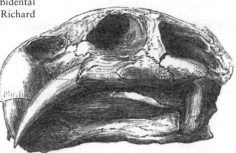

FIGURE 46 – The skull of Bain's "Bidental animal," named *Dicynodon* by Richard Owen.

Astonished by the unusual find, the Society sent Bain a £20 reward for his "judicious course" of action. Encouraged by the warm reception of the distant scholars, Bain started shipping more fossils and acted as the man in the field for London's cadre of professional naturalists.

While the rest of Bain's fossils were en route from the colony, the task of describing the enigmatic cranium fell to England's most eminent comparative anatomist, Richard Owen. The skull had belonged to some kind of reptile, Owen surmised, but it did not correspond to any known type of lizard, crocodile, or turtle. This extinct animal, which he named *Dicynodon* for the two enlarged fangs sticking out from the upper jaw, was a chimerical creature that consisted of both mammalian and reptilian parts. The ever-pious Owen could not help but think that the enigmatic fossil spoke to "a power transcending the trammels of the scientific system."

Dicynodon was only the first such creature to be described. South Africa proved to be unexpectedly fossil-rich, and Owen became the unofficial interpreter of the colony's prehistory. By the late 1850s Owen had seemingly everyone who ventured into the field, from laborers to visiting politicians, funneling fossils back to England for him to scrutinize. Even Prince Alfred, who visited South Africa in 1860, returned home with two more *Dicynodon* skulls for Owen to add to his expanding collection. His annoyance at the hubbub over Darwin's evolutionary theory aside, Owen was at the height of his scientific power.

The fossil flood from South Africa inundated Owen with a collection of "reptilian" skulls that also possessed mammalian traits, like differentiated kinds of teeth in different parts of the jaw. Owen celebrated these strange amalgamations of features by giving the new forms names like *Galesaurus* ("weasel reptile"), *Lycosaurus* ("wolf reptile"), and *Tigrisuchus* ("tiger crocodile"). Owen placed these creatures into a new group united by their mammal-like dentitions, the Theriodontia, and by

his estimate the fossils showed that some time in the distant past reptiles had begun to change into something more mammal-like.

Owen's interpretation was difficult for other naturalists to challenge. The senior scientist had a corner on the Karoo fossil market. If
a significant specimen was found it was often sent right to him. So pervasive was his reach that paleontologists who later visited the Karoo
lamented that almost all the best specimens had already been extracted
for Owen. This was made all the more intolerable by Owen's jealous
love for "his" fossils. Though kind to friends, Owen could be ruthless
with colleagues; he was a powerful and ornery figure aware of his own
brilliance.

Strangely, however, Owen's rival Thomas Henry Huxley was not
especially interested in these transitional forms, and on March 6, 1879,
Huxley stood before the Royal Society of London to present his views
on where the hairy, milk-producing group of vertebrates known as
"mammals" had come from. Though the origin of mammals was still
unresolved, Huxley argued, the identification of distinct evolutionary
series in the bird, whale, and horse lineages had shown that evolution
could be seen in the fossil record, after all. With the principle of evolution thus established clearly there must have once existed forms that
would connect mammals of modern aspect to their "protomammal"
antecedents.

At that time the "lowest" mammals known were the monotremes,
a group entirely represented by the duck-billed platypus and echidna.
The echidna had been described first, but it was the platypus that
caused a stir among naturalists when reports of it began to trickle back
from the far-flung British outpost of Australia at the turn of the nineteenth century. The platypus was covered in hair and secreted milk for
its young, yet it also had a ducklike bill and reproduced by laying eggs.
It seemed to defy the divinely imposed order of nature, and the naturalist Thomas Bewick remarked that the platypus "appears to possess a
three fold nature, that of a fish, a bird, and a quadruped, and is related
to nothing that we have hitherto seen."

So odd was the platypus that the first person to describe the remains
of one, George Shaw, confessed that it was "impossible not to entertain
some doubts as to the genuine nature of the animal, and to surmise that
there might have been practiced some arts of deception in its structure."
Such deception had often been practiced to fill the curiosity cabinets of
Europe, precursors to true museums, with "authentic" remains of
mythical creatures, but the acquisition of additional platypus specimens
soon convinced naturalists like Shaw that it was neither freak nor fake.

Though their relationship to other vertebrates was debated for a time, the monotremes were ultimately identified as a group of archaic mammals, and they could be distinguished from other mammals by their peculiar mode of reproduction: the young of monotremes hatched from reptilelike eggs. This contrasted with the types of early development seen in the other groups of mammals, the marsupials and placentals. Marsupials are the "pouched" mammals, like opossums and kangaroos, which give birth to tiny, underdeveloped young who crawl into their mother's pouch where they continue the rest of their early growth. The young of placental mammals, however, gestate for a longer period of time and are born more developed than newborn marsupials. There are other minute characteristics that distinguish these groups, but they are most easily separated from each other by the way in which they reproduce.

Despite their peculiar reptilelike traits, however, Huxley did not think that the monotremes indicated a direct evolutionary link between mammals and reptiles. Using the anatomy of pelvic bones as his guide, Huxley could find little that connected the form of the platypus to any known reptiles. Rather than mammals evolving *from* reptiles, then, Huxley proposed that both groups had gone through a few parallel stages of development from a salamanderlike amphibian.

> It seems to me that, in such a pelvis as *Salamandra*, we have an adequate representation of the type from which all the different modifications which we find in the higher Vertebrata may have taken their origins.

This wrapped the evolution of birds, reptiles, and mammals up nicely. Mammals and reptiles had evolved from salamanderlike amphibians, with birds being an offshoot of a later, dinosaurlike reptile.

The American paleontologist O. C. Marsh, who had recently astounded Huxley with his collection of toothed birds and miniature proto-horses during Huxley's visit to America, was inclined to agree with his British colleague. In an 1898 letter to the journal *Science*, Marsh expressed similar thoughts about the origin of mammals. Even two decades after Huxley's Royal Society address it seemed the fossils that would elucidate the origin of mammals were still missing. Marsh lamented:

> Too often in the past a discussion on the origin of mammals had seemed a little like the long philosophico-theological controversies in the Middle Ages about the exact position of the soul in the human body. No conclusion was reached, because, for one reason, there were no facts in the case that could settle the question, while the methods of investigation were not adapted to insure a satisfactory review.

Marsh insisted that the level of scientific discourse about mammalian origins had risen to a "higher plane" by the time of his writing, but hard evidence was still urgently needed. It was still unclear, for one thing, whether all mammals shared a common ancestry. The known types of mammals were so diverse that it would make the job of evolutionary scientists easier if the monotremes, the marsupials, and the placental mammals had separate origins.

Even if mammals had evolved multiple times, though, the identity of their ancestral stock remained mysterious. Reptiles were a popular choice, with some authorities pointing to the Karoo fossils Owen had described, but Marsh waved them away as instances of "parallel development." The possession of mammal-like teeth in the skull of a reptile, Marsh argued, could not be used as a clue to the true evolutionary relationship between groups because mammal-like dental arrangements had evolved multiple times in different societies of vertebrates. Using such a feature to trace evolutionary history would only throw scientists off the trail. Like his friend Huxley, Marsh thought that the earliest known fossil mammals were too unlike any known extinct reptiles to have evolved from them, and it was more probable that both reptiles and mammals had sprung from an unknown amphibian ancestor.

By the time Marsh wrote his letter to *Science*, though, many more "mammal-like reptiles" had been recognized. Numerous specimens had been uncovered as a side effect of copper mining in Russia, and the first of these were described at about the same time as Owen turned his attention to the Karoo fossils. In contrast to Owen, S. S. Kutorga, of the University of St. Petersburg, thought some of the forms he described, such as *Brithopus* and *Syodon*, were actually mammals rather than reptiles with mammal-like features.

The great "Bone Rush" of the late nineteenth century also turned up an array of bizarre new forms from North America. Named pelycosaurs by E. D. Cope, many of the North American fossils had been dug out of the rust-colored rock of Texas and seemed to show some resemblances to the Karoo fossils, such as the possession of differentiated canine and incisor teeth. But Cope also noted that the pelycosaurs, like the sail-backed *Dimetrodon*, closely resembled extinct amphibian relatives like the immense salamanderlike tetrapod *Eryops*. How they related to mammals, reptiles, amphibians, and even Owen's theriodonts was open for debate.

Other naturalists were just as puzzled by the collection of mammal-like creatures that had been accumulated by the twilight of the nineteenth century. Georg Baur, a young assistant to O. C. Marsh who was so committed to his research that he literally worked himself to death in 1898, considered Cope's fossils to be a collateral branch that only shared a common ancestor with the earliest mammals. Owen's theriodonts were either ancestral or very close to the earliest mammals, Baur hypothesized, yet they still did not create the graded series paleontologists were hoping to find. The British paleontologist H. G. Seeley was similarly frustrated. The fossils from North America and the Karoo possessed amphibian, reptilian, and mammalian characteristics. These animals were obviously close to the junctures where all three of these vertebrate groups split, but eventually Seeley, too, decided that the ancestors of mammals would be found among the early tetrapods of the Devonian or even further back in geologic time.

But the Scottish paleontologist Robert Broom though otherwise. Broom, who is more often remembered for his studies of early humans, spent decades studying the friable fossils of the Karoo desert, and in a 1915 review of mammalian origins Broom concluded that the features shared by both mammals and "mammal-like reptiles" indicated a close relationship between them, not an aberrant evolutionary pathway. Their bearing on the origin of mammals was "beyond question." According to Broom the pelycosaurs had given rise to the theriodonts from which the ancestors of mammals had evolved.

Broom initially proposed that a group of squat, weasel-like animals called cynodonts were the theriodont ancestors of mammals. These were relatively small creatures that held their limbs in a more vertical position under their bodies than their earlier relatives, and though early synapsids had only incisorlike and caninelike teeth, the cynodonts had the full complement of incisors, canines, and molars.

If Broom was correct, the root of the mammal family tree was not to be found among the early tetrapods of the Devonian, but among the anomalous "reptiles" of the Permian (which lasted from about 295 million years ago to 251 million years ago). As the next generation of paleontologists began to study these fossils in more detail they realized that the most powerful evidence of the relationship between them and the first mammals was among the most subtle. The general outline of synapsid evolution has changed quite a bit since the time of Owen and Cope. It all started with an egg.

The evolution of the amniotic egg—a self-contained pond, surrounded by an outer shell for developing vertebrates to grow in—around 330 million years ago allowed vertebrates to move away from the water and proliferate into a variety of forms. Among this radiation of amniotes were the early members of a group called the Synapsida, such as the 306-million-year-old *Archaeothyris*. If we could see *Archaeothyris* alive today we might be tempted to call it a lizard, but its skeleton exhibited some telltale characteristics that distinguish it from true reptiles. One of the most important was the presence of a single opening behind the eye socket in the skull called the temporal fenestra. This single opening, used as a site for jaw muscle attachments, is a defining feature of synapsids and is seen in members of this group to this day. (Even you and I have modified versions of this structure.) The earliest reptiles did not have this opening, and many of the later reptiles either had two temporal fenestrae or descended from ancestors with two temporal fenestrae.[43] The presence, position, and abundance of this major skull opening provides a quick and dirty way to separate early relatives of mammals from reptiles.

From early synapsids like *Archaeothyris* evolved other forms like the approximately 280-million-year-old *Eothyris*, another mock lizard with two pairs of large, canine-type teeth sticking out of its upper jaw. This differentiation in teeth would be further selected for among synapsids, as seen in the pelycosaurs.[44] These creatures, such as *Dimetrodon*, flourished during the early Permian, and included an array of forms from sail-backed predators to massive herbivores with comically small heads, such as *Cotylorhynchus*. Most, however, died out long before the massive extinction that rocked the planet at the era's close. While the relationships among the pelycosaurs are currently undergoing revision, members of a particular subgroup called sphenacodontians closely resemble a lineage of later synapsids with even more specialized features, the therapsids.

The therapsids are distinguishable from their pelycosaur ancestors by having larger temporal fenestrae, limbs that were held more verti-

FIGURE 48 – A restoration of *Dimetrodon*, an apex predator during the early part of the Permian. Despite its appearance, it was a synapsid and more closely related to us than to reptiles.

cally beneath their bodies, and distinct incisor, canine, and molar teeth. Although the exact details of their origins from the sphenacodontians are still being worked out, the earliest known therapsids evolved during the Middle Permian around 267 million years ago and quickly diversified into several different groups that replaced the more archaic pelycosaurs.[45]

Some of the earliest therapsids were the carnivorous biarmosuchians, predators that were clearly slightly modified versions of the sphenacodontian pelycosaurs. Their altered body plan made them more efficient hunters, however, and it was from creatures like this that other types of strange therapsids evolved. There was a mixed group of herbivores and carnivores known as the dinocephalians ("terrible heads"), the herbivorous anomodonts to which Owen's *Dicynodon* belonged, and the mostly carnivorous group known as the theriodonts, which included the terrible, saber-fanged gorgonopsians. As a whole, these therapsid groups evolved a wider diversity of forms than that seen in their pelycosaur forebears: the anomodont *Suminia* was one of the first vertebrates to live in the trees, herbivorous dinocephalians like *Moschops* had stout skulls reinforced with bone for literal head-to-head competition; and large gorgonopsians like *Inostrancevia* stripped the flesh off their prey with an advanced set of dental cutlery. The middle of the Permian was the heyday of synapsids, but many of these groups would soon be wiped out.

About 251 million years ago the earth suffered the worst mass extinction in its history. Most synapsid groups present at the time were mowed down by the catastrophe. Even so, some synapsids persisted through the event, and among the survivors were the cynodonts.

Cynodonts were the small therapsids that Broom had initially proposed as mammal ancestors, and even though they evolved shortly before the Permian extinction they somehow survived it. They underwent their own diversification in the wake of the extinction and, while most of the cynodonts faded into oblivion, at least one group among them was ancestral to the earliest mammals. By the time the earliest "true" mammals evolved during the early Jurassic, however, dinosaurs had already taken over as the dominant large vertebrates on land. Mammals started off as small, shrewlike creatures, and most would never grow larger than a house cat during the Mesozoic. The first mammals had evolved in a world overrun by dinosaurs, and the so-called "Age of Mammals" only began after non-avian dinosaurs became extinct. Indeed, descendants of the earliest synapsids survive to this day; you and I belong to the group. We are quite different from many of our early synapsid ancestors, but we retain ancient traits.

If we were to try to construct a pathway from sphenacodontian pelycosaurs through therapsids to the cynodonts and then modern mammals, the development of the mammalian inner ear might initially appear to be an insurmountable obstacle. How could the delicate association of miniscule bones (the incus, malleus, and stapes) that transmit sound from the outside world to our inner ear have evolved?

The bones of the mammalian inner ear did not appear out of nowhere. Mammals have only a single lower jaw bone, the dentary, but many early synapsids had multiple jaw bones that articulated with the back of the upper jaw in a way very different from that seen in modern mammals. These "extra" bones in the lower jaws of early synapsids would eventually become the components of the mammalian inner ear.

This connection between the mammalian jaw and ear had been known since the time Bain was digging for fossils in the Karoo desert. In 1837, the German anatomist C. B. Reichert noticed that during the embryonic development of a pig fetus, its lower jaw forms from a more flexible precursor called Meckel's cartilage. The formation of the lower jaw bone occurs through the transformation of part of the cartilage into bone, yet not all of the cartilage goes into the lower jaw. Some of the posterior part of the cartilage turns to bone and migrates into the inner ear, becoming the auditory bone called the "malleus." Through development, ear and jaw were connected.

Given that Reichert published this finding more than twenty years before the idea of evolution hit the scientific mainstream, however, it was not immediately recognized as relevant to the origins of mammals. It was the later accumulation of fossil evidence that would cause paleontologists to look back at and test what Reichert's observations had hinted at.

The story of the mammalian ear actually began long before the first synapsids or even the first amniotes. Tens of millions of years earlier, when there were no vertebrates on land, part of the gill arch of fishes was used to support the braincase of the skull. This bone was called the hyomandibular, and by the time the first tetrapods had evolved this bracing structure had been modified to assist in multiple functions. In the early tetrapods *Acanthostega*, for example, the hyomandibular was a small but stout bone that offered support to the skull, may have played a role in respiration (as it does in living lungfish), and was placed right next to the external ear opening on the head. This bone was in just the right position to conduct sound waves from the outside world to the skull. It had changed so much that it was given a new name, the stapes, and it is one of the chief components of the mammalian inner ear.

Early synapsids and pelycosaurs inherited stapes from their amniote ancestors, but they lacked the other ears bones seen in their later relatives. Skull shapes had changed quite a bit at this point, though, and the sound-conducting stapes was in direct contact with a bone called the quadrate in the upper jaw of creatures like *Dimetrodon*. This is important, as the quadrate bone was the part of the skull that the lower jaw articulated with, meaning that synapsids with this arrangement could hear through their jaws.

In mammals the quadrate bone is known by another name, the incus, but this auditory ossicle did not immediately detach from the lower jaw and migrate into the inner ear. In many therapsids, what we call the incus and the stapes were still connected to the lower jaw, and the lower jaw itself was undergoing a major reorganization. By setting up a sequence of skulls running from the pelycosaurs through the therapsids to the cynodonts, it could be seen that the dentary bone makes up more and more of the lower jaw as we look from the geologically older forms to the younger ones. In some cynodonts, in fact, the rear portion of the dentary even extends upward and backward to create a second contact with the upper jaw between a bone behind the dentary called the surangular and a bone of the upper jaw called the squamosal. The evolution of this second joint not only strengthened the jaw (with the force of chewing borne on one lower jaw bone), but allowed some lineages of cynodonts to switch between jaw joints.

This shift in jaw joint construction was essential to the later development of the inner ear. As the synapsids evolved, the muscles used in chewing became increasingly attached to the dentary, thus selecting for a larger dentary and smaller post-dentary bones. Even so, the bones behind the dentary in cynodonts still played a minor role in jaw support and movement in species where the articulation between the dentary and the squamosal bone of the skull had not been fully established. In species where the new jaw articulation was more firmly in place, however, the post-dentary bones were relieved from their roles in jaw movement, thus allowing them to be further coopted for hearing. One particular genus of cynodont, *Diarthrognathus* (roughly, "two-jointed jaw"), is virtually caught in the act of this transition, and in such cynodonts the dentary extends all the way back to contact the squamosal, successfully having switched to the kind of jaw joint seen in all mammals.

But even by the time the cynodonts had evolved a mammal-type lower jaw, they did not have inner ears like modern mammals. The forerunners to the tiny bones in our own ears had been shrunk down from parts of the more reptilelike lower jaw possessed by early synapsids, but they had not yet migrated inside the skull into the inner ear. Instead they became associated with the modified version of the angular bone that formed a ring that housed the tympanic membrane, a bit of soft tissue that acted like a drumskin to take in sound and transmit it to the ear bones. This ring fit into a notch that was positioned just behind the dentary in the lower jaw.

A recently described early mammal provides near-perfect evidence of this arrangement. In 2007 Luo Zhe-Xi and colleagues described the nearly complete specimen of an approximately 125-million-year-old mammal they called *Yanoconodon allini*. It was so well-preserved that the scientists were able to identify the malleus and incus in the fossil. These tiny bones were still connected to the lower jaw by a bit of ossi-fied Meckel's cartilage but had migrated farther from the jaw than in similar mammals that have been studied. In other words, what once were parts of the lower jaw were by now used for hearing and were only barely connected to the lower jaw. *Yanoconodon allini* confirmed what Reichert had glimpsed in

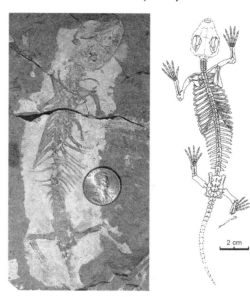

FIGURE 49 – The skeleton of *Yanoconodon allini* as preserved in the rock and compared to a restoration.

2 cm

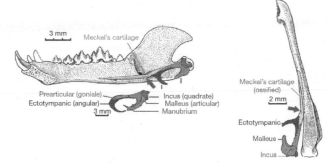

FIGURE 50 = The right lower jaw of *Yanoconodon allini* as seen from the inside and top. The ring on inner ear bones are connected to the lower jaw by way of a bar of ossified cartilage.

the development of pigs, and confirmed that mammals, as Stephen Jay Gould once put it, have "an earful of jaw."[46]

We owe another peculiar trait to our synapsid forebears: a secondary palate. It is what you are touching when you stick your tongue onto the top of your mouth, and it acts as a bony divider between your nose and mouth. This partition is important to an active lifestyle. Without it you would have to hold your breath while you ate your dinner.

Almost everyone, at one time or another, has been admonished by their parents to chew their food carefully. Careless masticatory behavior could cause an improperly chewed chunk of food to get lodged in the back of the throat, clogging the windpipe and putting us in mortal peril. We can blame this vulnerability on our early tetrapod ancestors. In early tetrapods the tubes used for breathing and swallowing food shared a common vestibule, the mouth. This arrangement meant that food could become clogged in the air-intake line with relative ease, and this is a danger all tetrapods have had to live with since. As if that were not enough, early tetrapods faced another challenge. They breathed by gulping air into their mouths, and in order to swallow their prey they would have to temporarily stop breathing.

This remained a problem even as the first amniotes evolved. Even though they had enlarged nasal openings, air still had to travel through the mouth to the windpipe, and air intake would be blocked while their mouths were full. This may not have presented much of a problem for small prey, but larger morsels posed more of a challenge. These creatures literally had to remember to breathe while they ate.

Many of the early synapsids encountered the same problem. The pelycosaurs and many of their therapsid descendants lacked the nasal/oral divider, yet it appears this feature evolved three separate times during synapsid history. This involved the extension of bones nestled near the nasal opening between the tooth rows of the upper jaw, and among other groups it evolved among the cynodonts.

Precisely why the secondary palate evolved and what function it might have initially had has been difficult to determine. Perhaps, by allowing synapsids to chew and swallow their food quickly, it allowed these animals to become endothermic (and hence more active). Then again, the secondary palate might have strengthened the upper jaw to withstand greater bite forces, allowing the synapsids to tackle a wider array of food. Whatever the reason for its evolution in several lines of synapsids, however, its presence in modern mammals is attributed to a quirk of history.

As Broom had hypothesized, the ancestors of the first true mammals were to be found among the cynodonts, the small synapsids who were among the survivors of the worst mass extinction ever suffered by life on earth. It was a catastrophe of unimaginable scale that no species could have prepared for. During such an event natural selection did not just work on the level of the individual animal, but on species and even entire groups of organisms. Some species already possessed features that allowed them to survive, while others suffered under the weight of such heavy selection pressures that they were extinguished entirely.[47] While the Permian world had been ruled by synapsids, they would not prevail again for more than 150 million years.

Extinction came quickly, in an event so catastrophic that paleontologist Michael Benton dubbed it "when life nearly died." Nearly seventy percent of all terrestrial vertebrates vanished, while over ninety-five percent of the known marine fossil species disappeared forever. It was as close as evolution has ever come to having its "redo from start" button pressed.

This is a relatively new understanding. Until recently it was believed that the Permian mass extinction occurred in an orderly manner in which each species dropped out of "life's race" one by one. This was the classic uniformitarian view that had been supposed since Darwin's time, but paleontologists have recently learned that large-scale catastrophes can happen. What we have come to find in the latest Permian and earliest Triassic rocks is that life on land and in the sea was almost wiped out virtually overnight.

It is often easier to understand the evolution of a species than its extinction. Every species carries a record, albeit a partial one, of its evolution in its biology, but how can we account for the disappearance of a species? Contrary to the beliefs of some early twentieth-century naturalists, extinction is not the result of a species reaching "old age" and disappearing at the end of the time natural forces have allotted them. Instead, mass extinctions are the effect of ecological interactions that cannot be observed by humans directly. We can see the damage, the wounds, but identifying the implements used during the massacre is extraordinarily difficult. To figure out what caused the Permian extinction, though, we need to look closely at the ancient crime scenes that record the tragedy.

The end-Permian mass extinction was most deeply felt in the seas. At that time the global ocean supported a wide array of coral reefs. Fish and the spiral-shelled ammonites congregated in the water above, while snails, single-hinged shelled mollusks called brachiopods, and other invertebrates practically swarmed on the bottom. These reef communities were not whittled away piecemeal but were eradicated in a geological instant. They were replaced by a wasteland inhabited by only a handful of species, which proliferated in the wake of the event. It was like looking at a forest recently leveled by fire; a small collection of fast-breeding "crisis" species lived where a more complex collection of organisms once thrived. Some groups, such as the last of the trilobites, were wiped out entirely, while others, such as brachiopods and the frond-like crinoids, persist to this day but never regained their previous levels of diversity.

Things were nearly as bad on land, which is best seen in the transition across the Permian–Triassic boundary in South Africa's Karoo desert. Around 251 million years ago this area was host to a collection of thirty-four genera of vertebrates, who lived along the horsetail-filled river margins and among the forests created by the treelike plant *Glossopteris*. Small cynodonts chased lizards and insects through the underbrush, while the far larger gorgonopsid predator *Rubigea* feasted upon the herbivorous *Dicynodon* and armored reptiles called pareiasaurs. The anomodont herbivores were much more abundant than their predators, however, and this slice of earth history is known as the *Dicynodon* Zone.

Much like the reefs offshore, the *Dicynodon* Zone represented a complete, complex ecosystem with various levels of interaction between its members. Near the top of the *Dicynodon* Zone, however, the abundance of vertebrates begins to dwindle, and on the other side of the boundary the "recovery" vertebrate community only boasts seventeen genera. Most of these are entirely different from the ones that had lived in the same place before. Approximately eighty-eight percent of the vertebrate genera disappeared during the transition, and the early Triassic fauna was made up of different members. The most numerous vertebrate was *Lystrosaurus*, a survivor from the end of the Permian and a relative of *Dicynodon*. It was also so numerous in the earliest Triassic strata that this complementary part of the series is called the *Lystrosaurus* Zone. It did not have as much to fear as its earlier relative, though. The large gorgonopsid predators did not survive. Other fossil localities in places like the Ural Mountains of Russia tell a similar story. Complex ecological webs were erased and replaced extremely abruptly. How could this have happened?

The world was not in stasis during the Permian. Toward its close, the large amount of CO_2 in the atmosphere caused the global climate to become much warmer and more arid. Lakes turned into seasonal floodplains and deserts began to spread. Even so, these changing conditions caused populations of organisms to continue to evolve, and while climate change can trigger extinction for some species, the gradual warming trend of the Permian is not enough to explain the mass extinction. Something else must have transpired.

There are several competing hypotheses for the trigger of the extinction. Some have argued that a massive asteroid or other extraterrestrial body must have struck the earth. Such an event is correlated with the mass extinction at the end of the Cretaceous, but so far no concrete evidence of such an impact has been found. Instead, it appears more likely that the planet's climate and atmosphere underwent some drastic changes that many organisms could not be adapted to. Just how such a shift might have happened, though, is open for debate. Some have argued that massive eruptions from the volcanoes known as the Siberian Traps would have choked the atmosphere with toxic gases. We know from violent eruptions in recent time, such as Mt. Saint Helens and Krakatoa, that eruptions can affect the global ecosystem by altering the atmosphere, but the eruptions alone do not seem to be sufficient for the scale of extinction that occurred.

Instead, the eruption of the Siberian Traps may have itself been a trigger. The intense volcanic activity may have caused the release of large amounts of carbon dioxide and methane stored beneath the earth's crust. The effects would have been disastrous. These greenhouse gases not only would have caused drastic changes to the global climate, but would have turned the seas into toxic, acidic pools in which most life would perish. Organisms on land did not suffer any less. The drastic change in the atmospheric content would have reduced the amount of available oxygen in the air. According to a 2003 paper by researchers Gregory Retallack, Roger Smith, and Peter Ward, after such an event the air in the Permian Karoo ecosystem would have been as thin as the air 14,000 feet above sea level today. This would make land-dwelling vertebrates susceptible to "nausea, headache, hypertension and pulmonary edema," similar to the maladies suffered by those who go up mountains too far, too fast.

If this hypothesis is correct, then the cause of the end-Permian extinction would have been more subtle than something like an asteroid impact. An asteroid or comet slamming into the earth would be like a gunshot, an instantaneous event that left horrendous damage in its aftermath. The release of large amounts of methane gas, instead, would strangle much of life on land by reducing the amount of available oxygen while toxifying the seas. The animals that survived would have been those who, simply by chance, had traits that allowed them to survive in the oxygen-deprived world. This hypothesis is not without problems, but so far it is the most consistent with the global pattern of extinction. Ideas about the cause of the crisis will continue to be debated, but we do know that some kinds of therapsids survived.

The recovery from the Permian mass extinction was relatively slow, but as the surviving groups continued to evolve reptiles began to outcompete the remaining synapsids. Early archosaurs, relatives of both dinosaurs and crocodylians, became the dominant vertebrates. Just like the therapsids, these creatures were independently evolving a posture with limbs held straight beneath their bodies. Some were even able to walk on two legs, and it was from one lineage of bipedal archosaurs that the first dinosaurs, similar to the small 230-million-year-old predator *Eoraptor*, evolved. Some Triassic archosaur groups would become extinct during yet another mass extinction that marked the end of the period, but dinosaurs would emerge in the Jurassic bigger and fiercer than anything that had come before.

The cynodonts continued to evolve alongside the dominant archosaurs, remaining small but persisting nonetheless. They were more efficient at obtaining and consuming food than their therocephalian ancestors, and at about the same time that the first dinosaurs evolved, the first mammals evolved from cynodonts. In these creatures the teeth were fully differentiated. They had an enlarged dentary bone that articulated with the skull and possessed a larger secondary palate. They were also probably covered in hair, even if they may have still laid eggs to reproduce. Just why the dinosaurs and other archosaurs dominated these synapsids is not fully known. (It has most often been thought of in terms of direct competition.) For the next 150 million years, mammals would not be larger than a small dog.

This is not to say that the dinosaurs were inherently "superior" to mammals. In fact, mammals survived and continued to evolve throughout the Mesozoic. They may have been small, but they carved out ecological niches small mammals occupy today. One of the most prolific groups during the Jurassic and Cretaceous, the multituberculates,

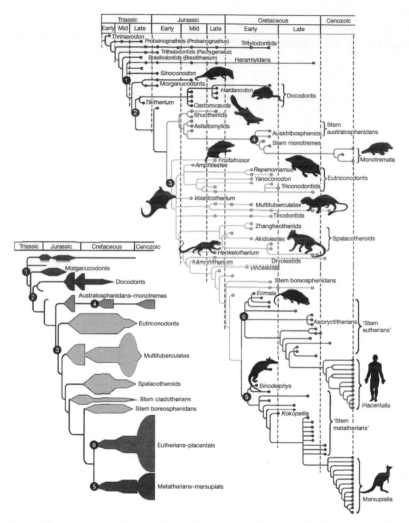

FIGURE 51 – An overview of mammal evolution during the Mesozoic. The larger diagram illustrates how many groups radiated and became extinct while the smaller summary shows the more general patterns of mammal diversity.

showed adaptations to digging in their skeletons. Others, like a few members of an early group known as docodonts, had beaverlike tails and adaptations similar to those seen in semi-aquatic mammals. There were even Mesozoic precursors to anteaters and flying squirrels among the early mammals of the time, and not all mammals avoided confrontation with dinosaurs. In 2005, a mammal named *Repenomanus*, from the Cretaceous of Mongolia, was found with baby dinosaur bones in its stomach.

All three major groups of mammals alive today also evolved during the reign of the dinosaurs. The egg-laying monotremes, the pouch-bearing marsupials, and the widespread placental mammals all trace their earliest ancestry deep into the Mesozoic. At first they might seem to represent three stages of recent mammal evolution, from odd egg-laying forms through marsupials to placentals, but the fossil record tells a different story.

Marsupial and placental mammals are more closely related to each other than either is to the monotremes. There is a vast evolutionary gulf between the two groups and the living archaic mammals like the duck-billed platypus. While all living mammals shared a common ancestor during the Early Jurassic, not all the descendant lineages survived until today. Living monotremes are the remnants of an early radiation of mammals, one that does not appear to have been as prolific as the mid-Jurassic explosion of multituberculates and other groups. The common ancestor of the marsupial and placental mammals, however, did not evolve until the Early Cretaceous. Thus monotremes, marsupials, and placental mammals do not represent a stepwise pattern of direct evolution, but represent varying outcomes of mammalian evolution that can be traced back to different times.

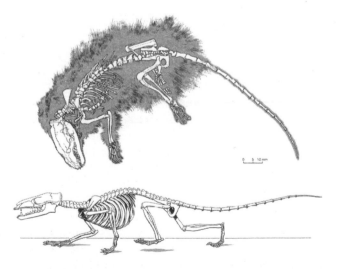

FIGURE 52 – A line drawing of the skeleton, preserved with fur, of *Eomaia*, as well as a restoration of the skeleton. This fossil mammal was a 125-million-year-old close relative of early placental mammals.

Then came another great extinction. Sixty-five million years ago, the impact of an asteroid in the area of what is today the Yucatan Peninsula threw scalding air, chunks of the earth's crust, and other debris into the atmosphere, triggering earthquakes and tidal waves in the immediate area of impact. From a distance it would have looked like the detonation of an extremely powerful bomb, but the worst effects of the event were felt over a longer span of time. The asteroid just happened to strike an area rich in sulfur, and the amount of the element the collision cast into the atmosphere caused the global temperature to drop up to ten degrees Celsius for several decades.[48]

The most famous victims of this extinction were the non-avian dinosaurs, but mammals had their numbers severely reduced by the extinction, too, and some groups disappeared altogether. Contrary to popular perception, mammals were hit almost as hard by the mass extinction as other groups of vertebrates, and only a few survived. The groups that fared better included the ancestors of living monotremes, marsupials, and placental mammals.

The extinction of the dinosaurs brought synapsids full circle. Nearly wiped out during the Permian extinction, the group diversified in a world ecologically dominated by dinosaurs. Many of their kind perished during the same extinction that wiped out the non-avian dinosaurs, but the survivors thrived just the way the archosaurs had. Things might have turned out differently had the asteroid missed. There is no reason to believe that if the extinction of the non-avian dinosaurs had been postponed or canceled, mammalian evolution would have proceeded in the same manner as we see it today. Despite what we might like to believe, the dinosaurs were not stifling some internal, directing potential that would have ultimately culminated in creatures like us. The mammalian fossil record as we understand it today is a story of contingency, and mass-extinction events weigh heavily on the way evolution unfolds.

In the aftermath of the Cretaceous extinction, new types of mammals flourished and were cut down by extinction. As Darwin rightly knew from his own study of fossil mammals, this pattern did not represent progress but the continued adaptation of the old into the new in the wake of ever-changing conditions. Among the most striking examples of how these contingencies and constraints molded vertebrates can be seen in the early evolution of whales, a transition that until very recently was one of the great mysteries of the fossil record.

As Monstrous as a Whale

"Scarcely anywhere in the animal kingdom do we see so many cases of the persistence of rudimentary and apparently useless organs, those marvellous and suggestive phenomena which at one time seemed hopeless enigmas, causing despair to those who tried to unravel their meaning, looked upon as mere will-of-the-wisps, but now eagerly welcomed as beacons of true light, casting illuminating beams upon the dark and otherwise impenetrable paths through which the organism has travelled on its way to reach the goal of its present condition of existence."

—WILLIAM HENRY FLOWER, "On Whales, Past and Present, and Their Probable Origin," 1883

In the summer of 1841 the German-born fossil collector Albert Koch unveiled a monster. Over thirty feet in length and fifteen feet high at the shoulder, the tusked wonder dwarfed the visitors to Koch's eclectic St. Louis Museum. He called it *Missourium theristrocaulodon*.

Koch had exhumed the remains of his star attraction not far from the lush banks of the Pomme de Terre River in Benton County, Missouri, the previous year. Among scraps of fossilized swamp moss and cypress trees were the bones of several individual animals, and Koch cobbled together their bones to construct a beast that seemed larger than life. Patrons flocked to the museum to marvel at its enormous tusks and tree-sized limbs.

Koch's *Missourium* proved popular enough that he decided to expand his audience. The beast was to go on an east coast tour, and one of the first stops was the Masonic Hall in the bustling port city of Philadelphia. Crowds came to marvel at what Koch sold as the ruler of the antediluvian world, but mixed among the curious members of the public were representatives of the city's prestigious community of naturalists.

Among them was the anatomist Paul Goddard, who immediately knew something was wrong. Koch's *Missourium* was not a new creature but was already familiar to naturalists as a mastodon, an extinct elephant given the name *Mammut americanum*. Even worse, Koch had erred in his reconstruction by adding extra ribs and vertebrae to inflate the stature of the already-gigantic proboscidean.

Koch's lack of academic training and his sensational promotion of his specimen did little to help him. Accusations and counter-accusations about Koch's lack of expertise and motives circulated through Philadelphia's scientific community, but Richard Harlan, another local anatomist and polymath, took the middle ground. After studying the skeleton himself Harlan could only conclude that Koch had simply made a few honest mistakes. Surely now that the errors had been exposed the extraneous bones would be removed.

If Koch agreed with Harlan's assessment, he did not let on. When the skeleton was erected in London's Egyptian Hall later the same year it appeared with every extra bone in place. Now, however, it had some competition. The scaly representatives of the recently described Dinosauria threatened to overshadow the "Missourium." As a consequence Koch played up the might and size of his ancient pachyderm. A broadside poster proclaimed:

> This unparalleled Gigantic remains, when its huge frame was clad with its peculiar fibrous integuments, and when moved by its appropriate muscles, was Monarch over all the Animal Creation; the Mammoth, and even the mighty Iguanodon may easily have crept between his legs.

Such fanfare did not deceive British naturalists. Even though the Missourium was greeted with enthusiasm by some members of the London scientific elite, in 1842 the anatomist Richard Owen pointed out the spare ribs and vertebrae his American colleagues had previously noticed. Koch passionately defended his reconstruction before the Geological Society, but the British scholars were not convinced.

In the wake of this controversy Koch took the skeleton on tour elsewhere in Europe, yet the London scientists had not entirely soured on Missourium. Despite Koch's overblown claims, it was still an impressive and valuable mastodon skeleton. When Koch stopped back in London in November 1843 at the end of the tour, the British Museum purchased it for £1,300. This was enough to enroll the Koch children in a school in Dresden, Germany, while the paleontologist and his wife traveled the European continent.

By 1844, however, Koch was itching to head back into the field to rebuild his collection. A fossil hunting trip through the United States would be just the thing, and as soon as Koch arrived in New England he started prospecting the local outcrops for choice specimens. Shells, shark teeth, and a few bones rewarded Koch's efforts, but what he was really after was another monster. The Yale professor Benjamin Silliman, a close friend of Koch, would be instrumental in providing him with one.

When Koch stopped in New Haven, Connecticut, to visit Silliman on August 17 he had his entire array of fossil treasures in tow. Silliman was impressed with what his friend had already accumulated, but he knew of another place where there were even more impressive bones to be found. In parts of southern Alabama, local residents had found the abundant remains of an enormous sea serpent, and Silliman knew a woman who could tell Koch where to chisel his own sea monster from the rock. They went off to meet her at once. If Koch could obtain these remains then he would surely have a new attraction.

With directions to the monster graveyard in hand, Koch continued through Pennsylvania, Ohio, Indiana, Kentucky, and Missouri, picking up fossils as he went. He finally reached Alabama on January 20, 1845, but the fossils he sought proved elusive. It would be another month before he would first catch sight of them, and they would not be in the place he expected. On February 14, Koch was on his way to meet an acquaintance in Macon County when he spotted an enormous, charred vertebra lying in a fireplace. It could only have come from the sea serpent.

When he asked about the scorched bone Koch was told that it had been used for nearly three years as a fireplace support. This was only one of the ways in which the plentiful fossils were regularly destroyed. During his travels Koch saw the gigantic vertebrae used to prop up a fence, as a cornerstone in a fireplace, and as a slave's pillow, and had even heard of a man who thought he could extract lime from the fossils by burning them. (All he got for his trouble were burnt bones that crumbled to pieces.) The bones were so numerous that in some fields they were destroyed because they interfered with cultivating the land. The widespread waste of the petrified treasures troubled Koch. As he wrote in his journal, it was a shame that so many fine specimens were "snatched from science by ignorance."[49]

Because these specimens were so lowly valued, however, Koch was able to procure specimens with relative ease. The day after an unproductive sojourn into the field, Koch headed toward Clarksville and spotted a few vertebrae outside an abandoned home. They had not yet been burnt to a crisp, and Koch was able to purchase one from the "ignorant but kindhearted" steward of the plantation.

Koch knew he had finally reached the resting place of the monsters, but he could not just dig anywhere he pleased. The best sites were all on plantations, and he would have to convince the landowners to let him scour the ancient marine deposits on their property.

By March 11, Koch still was starting to feel annoyed by his rotten luck. Not only had his sea serpent failed to emerge but a torrential downpour pinned him inside the home of his friend Colonel Washburn. Nevertheless, there was plenty of work to do. Koch busied himself by organizing some of the other fossils he had collected during his quest, but right in the middle of it all a mail rider arrived with a package for Koch's host.

Koch never had much patience when he was absorbed in his work, and when the postal carrier asked what he was doing Koch offered a curt reply to satisfy his curiosity and send him on his way. Fortuitously, though, the courier was no stranger to fossils. In fact, he had just heard that an immense stone shark, over ninety feet long, had just been found near the Washington County courthouse.

When Koch heard the mail rider's description he immediately knew that the bones could not have been from a shark, and he pressed the man for more details. Some vertebrae and ribs had been excavated, but they were too heavy to move, the mail rider said, so most of the bones were still in the ground. Word had gotten around about the find, though, and rumor had it that four vacationers from New York were intent on collecting the fossils. Koch would have to act fast, but even though the messenger offered to take him to the site directly the paleontologist declined. Washington County was a long way to ride for what might turn out to be just a big fish story. Instead, Koch told the mail rider to go to the site, find out everything he could, and return as quickly as possible. The trip was forty miles both ways.

The mail rider returned two days later, but he did not bring much more information about the skeleton. All he knew was that the party from New York had not yet acquired it. If Koch wanted to catch his monster he would have to take a gamble. Colonel Washburn lent him a horse, and the paleontologist was soon kicking up dust on his way to the old courthouse. On March 16, Koch arrived at the withering remains of the former county seat.

The only two people said to know anything about the "petrified shark," a young boy and a slave, were nowhere to be found, and Koch had to wait until the following day to meet them. It turned out that the fossil site, situated in the vicinity of the Sintabouge River, was even better than Koch had expected. The long chain of ribs and vertebrae was capped by a skull and most of the lower jaw.

Koch carefully teased the monster out of the rock, but as he did so news of another skeleton reached him. The remains of another gigantic animal rested just across the state line in Mississippi. With one sea serpent in hand Koch was free to try his luck on another, but after traveling more than forty miles all he found were a few crumbling vertebrae. He quickly returned to Alabama to pack up the Washington County fossils and spend a few more days out in the field.

To his surprise, Koch found a second skull on April 28. It would be a fine addition to his collection, but this time Koch had to contend with a swarm of onlookers. He wrote in his journal, "The many curious who come every day and who in their ignorance could break the whole thing cause me no small worry."

Koch's anxieties were realized the next day. As he packed up a few other fossils, one of the visitors took a more hands-on approach to learning and accidentally broke off a large part of the fossil's lower jaw. With his prize broken and his hired help in a drunken stupor, Koch had little recourse but to break up the skull and carry each piece back to his lodgings by himself.

Such frustrations aside, Koch had made a substantial haul. He had great plans to publicly unveil the impressive specimens he had collected and, on May 22, his fossils were sent by boat to New York. Koch himself arrived in the city just over a month later, but his sea monster was not there.

Koch's collection had departed from Mobile on the ship *Newark*, and its planned course was through the Gulf of Mexico, around Florida, and up the eastern coast of the United States. Unfortunately, the

FIGURE 53 – An illustration of Koch's *Hydrarchos* as it appeared on display.

ship never got very far. It had wrecked off Florida near Key West, and most of the cargo aboard was lost to the sea. Even when he was told that some crates had been saved Koch did not hold out much hope.

To Koch's surprise, the fossils were among the salvage and were sent to New York on the *Globe* free of charge. The fossilist anxiously awaited his shipment, grateful for the kindness of "those noble-minded men" who had saved his treasures.

When the crates arrived Koch was relieved to find that most of his collection was intact, and he quickly set about putting the sea monster bones together. By July 1845, Koch had assembled a 114-foot-long creature he called *Hydrarchos sillimani* in honor of his scholarly friend. It was an instant sensation. A commentator in the *New York Evangelist* wrote:

> Who knows but had he seen the Ark? Who knows but Noah had seen him from the window? Who knows but he may have visited Ararat? Who knows how many dead and wicked giants of old he had swallowed and fed upon? Perhaps when we touch his ribs, we are touching the residuum of some of Cain's descendants that perished in the deluge.

The *New York Dissector* also reveled in the magnificence of the beast. *Hydrarchos* was solid proof that the biblical Leviathan could not have been a more unimpressive whale or crocodile:

> The gorgeous portrait of Leviathan, in the matchless poetry of Job, has found its first conclusive prototype in this Hydrarchon—so striking so, indeed, to every scholar who will undertake a critical examination of the original language, as to completely supercede every animal heretofore proposed by commentators as the subject as the description . . . Indeed, we are confident it will ultimately be a point of unanimous opinion that the Leviathan is the apt and distinctive title which this re-discovered creature should permanently receive.

As far as many of Koch's patrons were concerned, the bones of *Hydrarchos* were the timeworn remains of a biblical monster whose modern day progeny were said to be spotted by sailors and visitors to New England beaches. Koch also marveled at the monster he had made, and he married the language of science and speculation in the promotional pamphlet for the exhibition. *Hydrarchos* was surely the most fearsome thing to have ever sculled the primeval seas.

The supposition that the *Hydrarchos* frequently skimmed the surface of the water, with its neck and head elevated, is not only taken from the fact, that it was compelled to rise for the purpose of breathing, but more so from the great strength and size of its head, which could, with the greatest ease, be maintained in an elevated position, when in the act of carrying in its jaws a Shark or a Saurier, while struggling for life, to free itself from the dreadful grasp with which it had been elevated from its native element, to serve as a morsel to this blood thirsty monarch of the waters.

But professional anatomists were less than reverent. If experts had been vexed by a few extra ribs in the Missourium, they nearly had fits over the errors in Koch's new curiosity.

The Harvard anatomist Jeffries Wyman was among the first to notice that the great serpent was not all it seemed. After inspecting it in New York he aptly deconstructed Koch's monster before the Boston Society of Natural History. The skull was too small for the body, the teeth looked more mammalian than reptilian, and the vertebral column was made up of bones from several individual animals. This latter point was indisputable for, as Wyman noted, each vertebra displayed a different degree of ossification. As an animal grows, cartilaginous parts of skeleton become ossified, and the chain of vertebrae from a single individual would never exhibit so many different states of ossification at one time. Even worse were the "paddles"; they were made up of fossil cephalopod shells resembling those of the pearly nautilus.

H. D. Rogers also had a look at the skeleton and reported that, among the assorted other fossils presented with the main attraction, he found parts of the ophidian amalgamation's inner ear. These fragments most closely resembled the same bones in the inner ear of whales. No reptile had anything that came close. The facts were clear. Either Koch had purposefully manufactured a fraud or had proven himself so incompetent that he was seeing sea monsters where there were none.

As with the Missourium, Koch was not inclined to change the mount. Several months after its New York appearance *Hydrarchos* resurfaced in Boston, with every scrap of bone in place, at same time that the British geologist Charles Lyell was visiting the city. Lyell was plagued daily with questions about sea monsters and *Hydrarchos*, but he did not share the public's enthusiasm. Lyell sent a letter to Professor Silliman expressing his opinion that *Hydrarchos* was nothing but a humbug.

Silliman apparently had his own reservations about having his name attached to the beast. Although he cautiously defended Koch's mount for several years, he suggested that Koch give credit where credit was due and change the name of the beast to honor Richard Harlan, who had described what he thought were the remains of a gigantic marine reptile the decade before. Koch acquiesced to his friend's wishes by rechristening his monster *Hydrarchos harlani*, and since Harlan had died two years earlier he could neither welcome nor object to the change.

All of this hullabaloo did not slow Koch down. After appearing in Baltimore, Philadelphia, and upstate New York, *Hydrarchos* was packed up and readied to be shipped across the Atlantic on May 4, 1846. A grand European tour was launched, including stops in Dresden and Berlin, and the crowds of continental Europe were just as enamored with the skeleton as the people of the United States. The exhibition of Hydrarchos in July 1847 in Berlin so impressed King Friedrich Wilhelm IV of Prussia that he purchased the skeleton for the city's Royal Anatomical Museum.[50]

Naturalists were still puzzled by the combination of reptilian and mammalian characteristics seen in the animal. Either Koch had made one of the most astounding fossil discoveries ever or certain academics were being played for fools. What might have been otherwise regarded as an unfortunate accident provided the conclusive evidence to resolve the issue.

While entertaining some naturalists in the Berlin museum, the German physiologist Jean Mueller was asked whether *Hydrarchos* was really a reptile or a mammal. Mueller pulled out the temporal bone from the skull to explain its relevance to this question, but to his horror the fragment slipped from his grip and shattered on the floor. When the unnerved scientists gathered all the fossil fragments, however, they noticed that the bone had been broken open in such a way as to reveal the characteristic structure of the inner ear. There was only one other kind of creature with an inner ear that matched: a whale. The great reptile was really a mammal.

Curiously, the controversy over whether the giant bones of this oceangoing monster belonged to a reptile or a mammal had been played out once before: in 1832, twelve years before Koch set out to find his sea monster, an Arkansas man named Judge Bry sent a package to the American Philosophical Society in Philadelphia. While farmers in Alabama might have considered the large vertebrae that littered their fields a nuisance, Bry thought they might be of some scientific interest. When part of a hill collapsed on his property and exposed a string of twenty-eight of the circular bones, he decided to send one to scholars in Philadelphia. No one quite knew what to make of them. Some of the sediment attached to the bone contained small shells that showed that the large creature had once lived in an ancient sea, but little more could be said with any certainty.

Bry's donation was soon matched, and even exceeded, by that of Judge John Creagh from Alabama. Creagh, who lived in the area Koch would later search for his sea monster, had found vertebrae and other fragments while blasting on his property and, like his Arkansas colleague, sent off a few samples to the Philadelphia society. This time the bones were reviewed by Richard Harlan. The fossils were unlike any Harlan had seen before, but he needed more of the skeleton to determine just what sort of animal they belonged to. Working with fellow naturalist J. P. Wetherill, Harlan contacted Creagh and asked him to obtain as many fossils of the same animal as possible, especially the skull and jaws.

Creagh was happy to comply. Harlan soon received parts of the skull, jaws, limbs, ribs, and backbone of the enigmatic creature. Given that both Creagh and Bry said they had seen intact vertebral columns in excess of 100 feet in length, the living creature must have been one of the largest vertebrates ever to have lived. But what kind of animal was it?

Comparative anatomy was the key to unlocking the mystery. After comparing the fossils to those of other animals in 1834, Harlan determined that the bones were most similar to those of extinct marine reptiles such as the long-neck plesiosaurs and streamlined ichthyosaurs. Even so, Harlan was cautious. Perhaps further discoveries would reveal unexpected attributes of the creature, but for now enough about it had become known to tentatively ascribe it a name.

> If future discoveries of the remaining portions of this skeleton, should confirm the indication above pointed out, we may suppose the genus to which it belonged will take the name, not inappropriately, of *Basilosaurus*.

The lack of a complete skeleton was problematic, but more troubling was a portion of the jaw containing several teeth that differed in size and shape. This condition of different types of teeth in the jaw is known as heterodonty, a characteristic of mammals but not most reptiles.[51] Why did the largest fossil reptile that ever lived have mammal-like teeth? The question was certainly perplexing, and it would play a major role in a debate that was beginning to brew among London's scientific elite.

Prior to 1838, many geologists thought that there was a temporal dividing line between the "Age of Reptiles," during which the immense dinosaurs roamed, and the subsequent "Age of Mammals." Mammals could not have survived among the marauding reptiles, it was thought, and instead represented the next progressive step in the order of nature in the succeeding era. It came as a surprise, then, that several sets of mammalian jaws had been found between 1812 and 1818 in the town of Stonesfield from the same strata that produced dinosaurs. No less an authority than Georges Cuvier had identified them as belonging to opossumlike mammals, but the controversy did not fully erupt until similar jaws were found during the 1830s.

The French geologist Constant Prevost had argued that the jaws were out of proper geologic context, while the Swiss zoologist Louis Agassiz thought that they most closely resembled the jaws of fish.[52] The French anatomist Geoffroy Saint-Hilaire took a different view, welcoming the idea that mammals had lived alongside dinosaurs but arguing that since the jaws were from marsupials (which he did not consider true mammals) they were not sufficient to confirm the presence of mammals during the Age of Reptiles.

The alternate hypothesis that received the most attention, however, was that of the French anatomist H. M. D. de Blainville. While he admitted that the remains of mammals might eventually be found from the same strata as dinosaurs, Blainville was confident that the jaws belonged to reptiles. To support this argument he referred to *Basilosaurus*, a reptile with mammal-like teeth, but this was a risky move. Richard Owen, who defended the mammalian identification of the Stonesfield jaws, cautioned Blainville against using *Basilosaurus* to support his case. A detailed study of the teeth of *Basilosaurus* had yet to be made.

By a lucky coincidence Harlan traveled to London in 1839 to present *Basilosaurus* to some of the leading paleontologists and anatomists of the day. It would be Owen, a rising star in the academic community, who would get the privilege of studying it. Owen carefully scrutinized every bone, and he even received permission to slice into the teeth to study their microscopic structure. Owen's attention to such tiny details would ultimately settle the case. *Basilosaurus* did share some traits with marine reptiles, but this was only a superficial case of convergence since the creature had also lived in the sea. The overall constellation of traits, including double-rooted teeth, unquestionably identified *Basilosaurus* as a mammal. With Harlan's permission Owen rechristened the creature *Zeuglodon*.

Those who had some difficulty pronouncing the new moniker sometimes bastardized it to "Zygodon," and this was the name by which Koch called it in his travel journal. He had known all along

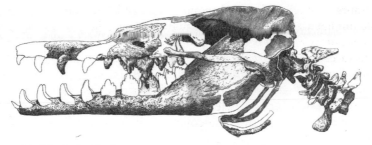

FIGURE 54 – The skull of *Basilosaurus*.

that he was on the trail of an already-known marine mammal. Indeed, when Koch produced a second, smaller skeleton from *Basilosaurus* bones in 1849 he called it *Zeuglodon macrospondylus*. It more closely resembled the short-necked, long-bodied skeleton geologist S. B. Buckley had dug out of Judge Creagh's property in 1841. Koch had ignored the research of his peers so that he could create something even more magnificent.

Despite the controversy and confusion Koch's first sea monster had caused, the true identity of the enormous bones from Alabama had been resolved. The teeth confirmed that it was a mammal, and the telltale inner ear bones noticed by H. D. Rogers and Jean Mueller classified *Basilosaurus*, its proper name given Harlan's priority, as a whale.

Not long after the true identity of *Basilosaurus* became known, however, Charles Darwin's evolutionary vision raised questions about the pedigree of the sharp-toothed whale. What it had evolved from or was ancestral to (if anything) was a mystery. Indeed, the origin and evolution of whales was inscrutable, and the response Darwin received to a thought experiment in the first edition of *On the Origin of Species* made him wary of even speculating on the subject. Based upon observations of black bears made by the explorer Samuel Hearne in 1795, Darwin supposed that, given enough time and the right selection pressure, a bear could evolve into something whalelike.

> In North America the black bear was seen by Hearne swimming for hours with widely open mouth, thus catching, like a whale, insects in the water. Even in so extreme a case as this, if the supply of insects were constant, and if better adapted competitors did not already exist in the country, I can see no difficulty in a race of bears being rendered, by natural selection, more and more aquatic in their structure and habits, with larger and larger mouths, till a creature was produced as monstrous as a whale.

Darwin was widely ridiculed for this passage. Critics often took it to mean that he was proposing bears as direct ancestors to whales. Darwin had done no such thing, but even though he privately stuck to his proposition the jeering caused him to modify the passage in subsequent editions of the book. Yet praise for the hypothesis came from an unexpected source. As related to Charles Lyell in an 1859 letter, Darwin was surprised to gain support on the subject from his "bitter & sneering" adversary Richard Owen:

> Lastly I thanked him for Bear & Whale criticism, & said I had struck it out. —"Oh have you, well I was more struck with this than any other passage; you little know of the remarkable & essential relationship between bears & whales".—
>
> I am to send him the reference, & by Jove I believe he thinks a sort of Bear was the grandpapa of Whales!

Darwin left the subject of whales alone in most of the following editions of *On the Origin of Species*, but while preparing the sixth edition he decided to include a small note about *Basilosaurus*. Writing to his staunch advocate T. H. Huxley in 1871, Darwin asked whether the ancient whale might represent a transitional form. Huxley replied that there could be little doubt that *Basilosaurus* provided clues as to the ancestry of whales.

Huxley had reviewed the evidence for such a connection a year before in his presidential address to the Geological Society of London. While he could not call it a definite ancestor of living whales, Huxley thought that *Basilosaurus* at least represented the type of animal that linked whales to their terrestrial ancestors. If this was true then it seemed probable that whales had evolved from some sort of terrestrial carnivorous mammal. Another extinct whale called *Squalodon*, a fossil dolphin with a wicked smile full of triangular teeth, also hinted that whales had evolved from carnivorous mammals. Like *Basilosaurus*, though, *Squalodon* was fully aquatic and provided few clues to the specific stock from which whales arose. Together these fossil whales hung in a kind of scientific limbo, waiting for some future discovery to more forcefully connect them with their land-dwelling ancestors.

This general lack of other transitional forms stirred debate about the relationship of *Basilosaurus* to other mammals (the biologist D'Arcy Thompson, for example, thought it was more closely related to seals than to whales), but the anatomist William Henry Flower was certain it was relevant to whale evolution. In an 1883 lecture reviewing all that was known of whales, collectively called cetaceans, Flower reviewed the common suggestion that *Basilosaurus* had been preceded by a seal-like stage in evolution, and this as-yet-undiscovered type would exist in the gap between *Basilosaurus* and the land-dwelling carnivores authorities like Huxley thought gave rise to whales.

But this hypothetical sequence was not without its problems. Flower noted that seals and sea lions used their limbs to propel themselves through the water, while whales lost their hind limbs and swam by oscillations of their tail. He could not imagine that early cetaceans used their limbs to swim and then switched to tail-only propulsion at some later point. The semi-aquatic otters and beavers were better alternative models for the earliest terrestrial ancestors of whales. If the early ancestors of whales had large, broad tails it could explain why they evolved such a unique mode of swimming.

Flower could not unequivocally support living mammalian carnivores as the best models for early whale ancestors, either. Ungulates, or hoofed mammals, shared some intriguing skeletal similarities with whales. To Flower the skull of *Basilosaurus* had more in common with ancient "pig-like Ungulates" than seals, thus giving the common name for the porpoise, "sea-hog," a ring of truth. If ancient omnivorous ungulates could eventually be found, Flower reasoned, it would be likely that at least some would be good candidates for early whale ancestors. Flower hastened to add that his ideas were still speculative, but he ended his lecture by envisioning a hypothetical cetacean ancestor easing itself into the shallows.

> We may conclude by picturing to ourselves some primitive generalized, marsh-haunting animals with scanty covering of hair like the modern hippopotamus, but with broad, swimming tails and short limbs, omnivorous in their mode of feeding, probably combining water plants with mussels, worms, and freshwater crustaceans, gradually becoming more and more adapted to fill the void place ready for them on the aquatic side of the borderland on which they dwelt, and so by degree being modified into dolphin-like creatures inhabiting lakes and rivers, and ultimately finding their way into the ocean.

The fossil remains of such a creature remained elusive. By the turn of the twentieth century the oldest fossil whales were still represented by *Basilosaurus* and similar forms like *Dorudon* and *Protocetus*, all of which were fully aquatic. As E. D. Cope admitted in an 1890 review of whales, "The order Cetacea is one of those of whose origin we have no definite knowledge." This state of affairs continued for decades.

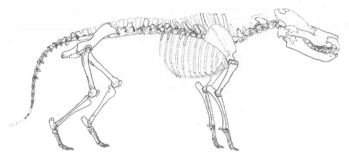

FIGURE 55 – A reconstruction of the mesonychid *Pachyaena*. It was once thought that whales evolved from these carnivorous mammals.

While revising the relationships of ancient meat-eating mammals in 1966, however, the evolutionary biologist Leigh van Valen was struck by the similarities between an extinct group of carnivores called mesonychids and the earliest known whales. Often called "wolves with hooves," mesonychids were medium- to large-sized predators with long, toothy snouts and toes tipped with hooves rather than sharp claws. They were major predators in the northern hemisphere from shortly after the demise of the dinosaurs until about thirty million years ago, and the shape of their teeth resembled those of whales like *Protocetus*.

The known mesonychid menagerie did not illustrate a finely graded transition from the land to the water, but they were the best candidates available for the long-lost ancestors of whales. Van Valen hypothesized that some of them may have been marsh dwellers that "were mollusk eaters that caught an occasional fish, the broadened phalanges [finger and toe bones] aiding them on damp surfaces." A population of mesonychids who lived in a marshy habitat, then, might have been enticed into the water by seafood. Once they had begun swimming for their supper, succeeding generations of the population would become more and more aquatically adapted until something "as monstrous as a whale" evolved.

Van Valen's mesonychid hypothesis sparked new interest in the origins of cetaceans, but there was still little to work with. Even the evolution of modern whales remained mysterious. The two living whale groups, the baleen whales (mysticetes, such as the humpback, blue, and right whales) and the toothed whales (odontocetes, such as dolphins, porpoises, and the orca) were so different that some researchers could not imagine how they could have evolved from a common ancestor. Perhaps each group had their own distinct terrestrial ancestors, with *Basilosaurus* being a third offshoot of whale evolution.

As if on cue, the discovery of a fossil mysticete with teeth, *Aetiocetus*, sunk the independent origins hypothesis. It confirmed that both odontocetes and mysticetes had evolved from toothed ancestors. The common ancestry of toothed and baleen whales was stabilized, but year after year rolled by without any sign of creatures recording the transition from land to the water. *Protocetus* was still the closest whale approximation to the presumed ancestral mesonychid type (often represented by the fifty-six-million-year-old *Sinonyx* from China).

This lack of early transitional forms led some scientists to doubt that transitional whales would ever be discovered. In a 1976 paper, Jere Lipps and Edward Mitchell noted that archaeocetes (early whales that are not odontocetes or mysticetes) appeared abruptly in the fossil record. There was no explanation for why this should be, and whale evolution did not seem to fit the traditional "slow and steady" model. This had long been a bugbear of evolutionary paleontology, all the way back to when Darwin was formulating his evolutionary theory. While the authors recognized that the lack of transitional forms might mean that paleontologists had not been looking in the right places, Lipps and Mitchell thought it more likely that whale evolution happened so fast that the fossil record had preserved little sign of it.

A startling discovery made in the arid sands of Pakistan, announced by University of Michigan paleontologists Philip Gingerich and Donald Russell in 1981, finally delivered the transitional form scientists had been hoping for. In freshwater sediments dating to fifty-three million years ago, the researchers recovered the fossils of an animal they called *Pakicetus inachus*. Little more than the back of the animal's skull had been recovered, yet it possessed a feature that would unmistakably connect it to cetaceans.

It is easy to identify modern whales by their tail flukes, blowholes, and blubbery hide, but these are all later adaptations to life in the sea which were not possessed by the earliest whales. The features that link all whales are less obvious, and one of the most important is found on the skull. Cetaceans, like many other mammals, have ear bones enclosed in a dome of bone on the underside of their skulls called the auditory bulla. Where whales differ is that the margin of the dome closest to the midline of their skulls, called the involucrum, is extremely thick, dense, and highly mineralized. This condition is called pachyosteosclerosis, and whales are the only mammals known to have such a heavily thickened involucrum. The skull of *Pakicetus* exhibited just this condition.

Even better, the teeth of *Pakicetus* were very similar to those of mesonychids. It appeared that van Valen had been right, and *Pakicetus* was just the sort of marsh-dwelling creature he had envisioned. The

fact that it was found in freshwater deposits and did not have specializations of the inner ear for underwater hearing showed that it was still
very early in the aquatic transition. Gingerich and Russell wrote,
"[Pakicetus is] an amphibious intermediate stage in the transition of
whales from land to sea. Postcranial remains will provide the best test
of this hypothesis." The scientists had every reason to be cautious, but
the fact that a transitional whale had been found was so stupendous
that full-body reconstructions of Pakicetus appeared in books, magazines, and on television. It was presented as a stumpy-legged, seal-like
creature, an animal caught between worlds that hinted that other such
creatures might yet be found from the Eocene rock of Pakistan.[53]

There were still things to learn about Basilosaurus, too. In 1896, the
paleontologist Charles Schuchert discovered a pair of hips and part of
a femur associated with a Basilosaurus skeleton. It was difficult to tell
whether the hips were attached to functioning limbs or were suspended
in the flesh of the body wall, but since some living whales have cartilaginous vestiges of hips and hind limbs inside their body it seemed fair
to assume that Basilosaurus, too, had lost its external hind limbs. There
would be no reason for the whale to have them if it was indeed fully
aquatic, but in 1990 Gingerich and several colleagues described numerous Basilosaurus hind limb fragments found in Egypt. No single leg
was entirely preserved, but the scientists had found enough bones to
almost fully reconstruct the small, three-toed hind limbs that would
have stuck out on the underside of Basilosaurus. Similar discoveries
revealed that other ancient whales like Takracetus and Gaviocetus had
external hind limbs, too. Perhaps these archaeocetes were not so far
removed from their terrestrial ancestors as had been thought.

Throughout the 1990s the skeletons of other more-or-less aquatically adapted archaeocetes were discovered at a dizzying pace, but a more
complete skeleton of Pakicetus proved elusive. The stubby, seal-like form
depicted in so many places began to make less and less sense as more
archaeocetes became known. It was not until 2001 that J. G. M. Thewissen and colleagues described the sought-after skeleton.

The skeleton of Pakicetus attocki, reconstructed next to that of its
smaller relative Ichthyolestes, was that of a wolflike animal. It was not
the slick, seal-like animal that had originally been envisioned. Together
with other recently discovered genera like Himalayacetus, Ambulocetus,
Remingtonocetus, Kutchicetus, Rodhocetus, and Maiacetus however, it
fits snugly within the graded series of archaeocetes that exquisitely document early whale evolution. Just as Huxley could not call Basilosaurus
a direct ancestor of living whales, though, the panoply of known archaeo-

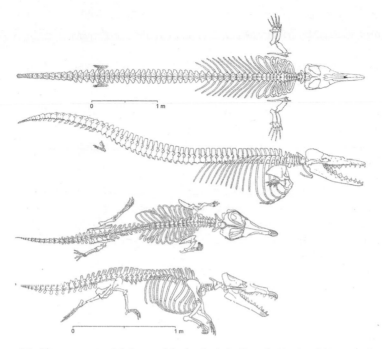

FIGURE 56 – The reconstructed skeletons of the fossil whales *Dorudon* (top) and *Maiacetus* (bottom). Together they represent two grades of whale evolution, with *Maiacetus* still being capable of moving around on land and *Dorudon* being fully aquatic.

cetes cannot be arranged in single evolutionary trajectory. It is not possible to draw a direct line of descent from *Pakicetus* through *Ambulocetus*, *Rodhocetus*, *Georgiacetus*, and *Basilosaurus* to modern whales. Instead, each genus represents a particular stage of whale evolution and together they illustrate how the entire transition took place.

The earliest known archaeocetes were creatures like the fifty-three-million-year-old *Pakicetus*. They looked as if they would have been more at home on land than in the water, and they probably got around lakes and rivers by doing the doggie paddle. A million years later there lived other archaeocetes like *Ambulocetus*, an early whale with a crocodilelike skull and large webbed feet. The long-snouted and otterlike remingtonocetids appeared next, with small forms like the forty-six-million-year-old *Kutchicetus* exemplifying the greater diversity archaeocetes were attaining. These early whales lived throughout nearshore environments, from saltwater marshes to the shallow sea nearby; not all were becoming adapted to life in the ocean.

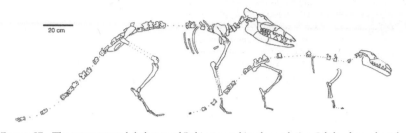

FIGURE 57 – The reconstructed skeletons of *Pakicetus* and its close relative, *Ichthyolestes*, based on more complete fossil material.

Living at about the same time as the remingtonocetids was another group of even more aquatically adapted whales, the protocetids. These forms, like *Rodhocetus*, were nearly entirely aquatic. In fact some later protocetids, like *Protocetus* and *Georgiacetus*, almost certainly lived their entire lives in the sea and dispersed to the shores of other continents. This shift allowed the fully aquatic whales to expand their ranges and diversify, and the sleeker basilosaurids like *Dorudon*, *Basilosaurus*, and *Zygorhiza* populated the warm seas of the late Eocene. These forms eventually died out, but not before giving rise to the early representatives of the two groups of whales alive today, the odontocetes and mysticetes. The early representatives of these groups appeared about thirty-three million years ago near the boundary between the Eocene and the Oligocene and ultimately gave rise to forms as diverse as the gigantic blue whale and the Yangtze River dolphin.[54]

The discovery of so many early fossil whales finally allowed researchers to reassess van Valen's mesonychid hypothesis, which primarily rested on dental evidence. The teeth of early archaeocetes and mesonychids were extremely similar, so much so that initially the teeth of the early whales *Gandakasia* and *Ichthyolestes* were thought to have belonged to mesonychids. This strengthened the mesonychid–archaeocete link, for if the fragmentary remains of one group could so easily be thought to belong to those of another it was probable they shared a close relationship.

Studies coming out of the field of molecular biology conflicted with the conclusions of the paleontologists, however. When the genes and amino acid sequences of living whales were compared to those of other mammals the results often said that whales were most closely related to artiodactyls, even-toed ungulates like antelope, pigs, and deer. Even more surprising was that comparisons of the proteins used to determine evolutionary relationships often placed whales *within* the Artiodactyla as the closest living relatives to hippos.

FIGURE 58 – The astragali (ankle bones) of several mammals. The astragali of the fossil ungulates *Phenacodus* and *Pachyaena* (a mesonychid) resemble each other in having only one "pulley." The ankle of *Pakicetus* and other early whales, however, share the specialized "double pulley" shape with the early artiodactyls *Diacodexis* and a living artiodactyl, the pig.

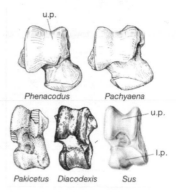

Phenacodus   Pachyaena

Pakicetus   Diacodexis   Sus

This conflict between the paleontological and molecular hypotheses seemed intractable. Mesonychids could not be studied by molecular biologists because they were extinct, and no skeletal features had been found to conclusively link the archaeocetes to ancient artiodactyls. Which were more reliable, teeth or genes? The conflict was not without hope of resolution. Many of the skeletons of the earliest archaeocetes were extremely fragmentary, and they were often missing the bones of the ankle and foot. One particular ankle bone, the astragalus, had the potential to settle the debate. In artiodactyls this bone has an immediately recognizable "double pulley" shape, a characteristic mesonychids did not share. If the astragalus of an early archaeocete could be found it would provide an important test for both hypotheses.

It was not until 2001 that archaeocetes possessing this bone were described, but the results were unmistakable. Archaeocetes had a "double pulley" astragalus, confirming that cetaceans had evolved from artiodactyls. Mesonychids could not longer be considered the ancestors of whales.

The group of artiodactyls that gave rise to whales, however, was still unknown. Hippos were the closest living relatives to whales but they and their extinct relatives (the anthracotheres) were already too specialized to represent the kind of animal whales evolved from. A better candidate was presented in 2007 when J. G. M. Thewissen and other collaborators announced that *Indohyus*, a small deerlike mammal

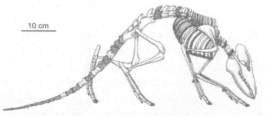

10 cm

FIGURE 59 – The reconstructed skeleton of *Indohyus*. A close relative of early whales, it may represent the kind of animals from which the first cetaceans evolved.

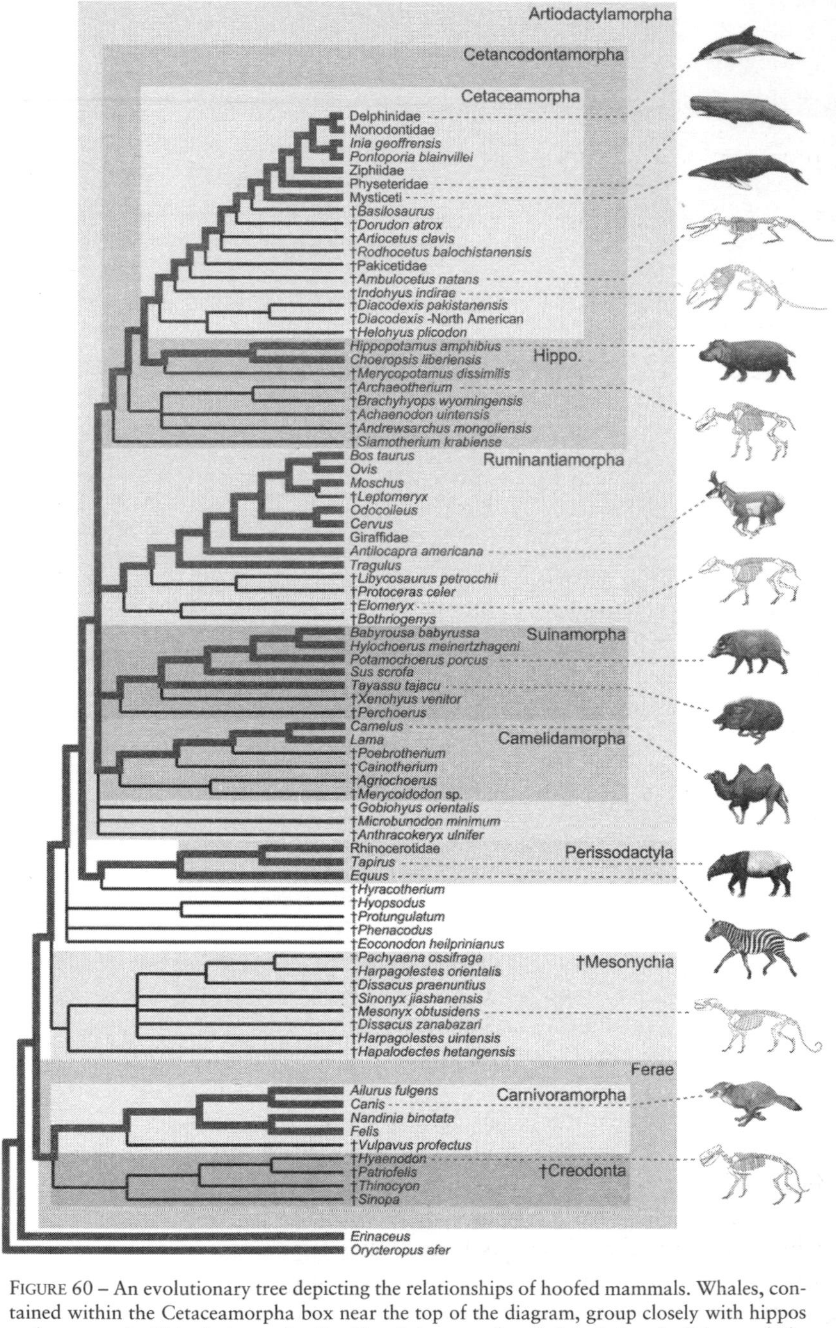

FIGURE 60 – An evolutionary tree depicting the relationships of hoofed mammals. Whales, contained within the Cetaceamorpha box near the top of the diagram, group closely with hippos and are nested within the larger family of hoofed mammals called Artiodactyls. (Mesonychids, once thought to be ancestors of whales, group most closely with other carnivorous mammals near the bottom of the diagram.)

belonging to a group of extinct artiodactyls called raoellids, was the closest relative to whales.

It was another accident that established this relationship. While preparing the underside of the skull of *Indohyus*, a student in Thewissen's lab broke off the section covering the inner ear. It was thick and highly mineralized, just like the bone in whale ears. Study of the rest of the skeleton also revealed that *Indohyus* had bones marked by a similar kind of thickening, an adaptation shared by mammals that spend a lot of time in the water. While it was too late to be a direct whale ancestor, *Indohyus* preserved just the sort of traits that the archaeocete ancestor would have possessed. When the fossil data was combined with genetic data by Jonathan Geisler and Jessica Theodor in 2009, a new whale family tree came to light. Raoellids like *Indohyus* were the closest relatives to whales, with hippos being the next closest relatives to both groups combined (as well as the closest living relatives to whales given the extinction of the raoellids). At last, whales could be firmly rooted in the mammal evolutionary tree.

Establishing the outline of cetacean evolution and their relationships to other mammals has been an important task, but the "archaeocete revolution" extends beyond taxonomic classification. The fossil remains of archaeocetes also document how the transition from land to water was actually effected. Whales were not the first vertebrates to become secondarily adapted to life in the sea, however, and in order to fully understand the unique origin of whales it is profitable to look at a much older group of animals. Two hundred and forty-five million years ago, while the ancient ancestors of artiodactyls inhabited the corners of a world ruled by dinosaurs, a peculiar group of reptiles became adapted to life in the water.

There were many types of marine reptiles in the past, all of which evolved from terrestrial ancestors, but they were shaped in different ways by historical contingencies. The long- and short-necked plesiosaurs, the lizardlike mosasaurs, the barrel-bodied placodonts, and sea turtles all became independently adapted to marine life. One particular group, however, looked like they could have been an early version of the dolphin prototype: the ichthyosaurs. Their long snouts and streamlined bodies made them look superficially similar to living marine mammals, but the convergence was not perfect.

One of the most obvious differences between dolphins and ichthyosaurs is that dolphins have flukes that are held horizontally while ichthyosaurs had caudal fins oriented vertically, meaning that dolphins swim by beating their tails up and down and ichthyosaurs by moving their tails from side to side. These differences have everything to do with how the early terrestrial ancestors of both groups moved. The ancestors of ichthyosaurs were reptiles that had a sprawling posture in which their bent legs were held out to the side. This means that when they walked they moved in a side-to-side motion and upon entering the water used their legs and tail in a similar way to swim. Eventually the tail became broader and flattened for propulsion, and their limbs became more useful for steering than swimming. This is just the type of condition seen in the "proto-ichthyosaurs" called icthyopterygians. The terrestrial ancestors of these animals are still unknown, but like *Ambulocetus* the ichthyopterygians are intermediate forms that represent the early stages of life in the water.

Small, early Triassic members of this group, like *Utatsusaurus* and *Chaohusaurus* had long bodies, stubby limbs, no dorsal fin, and just a bump for a caudal fin. This is known from body impressions of these animals from China and British Columbia, and they were similar in form to dogfish sharks. They probably swam much like eels do, and they may have been ambush predators that relied on quick bursts of speed to catch prey, as they would not be able to sustain swimming at high speeds for very long. From these creatures the first true ichthyosaurs evolved, typified by forms like the miniscule *Mixosaurus* and the enormous *Cymbospondylus*. These genera had even larger fins than their predecessors and swimming abilities more comparable to modern requiem sharks, like the tiger shark.

By the late Triassic ichthyosaurs had become more speedy and streamlined than any of their predecessors. These tunalike ichthyosaurs diversified and spread all over the globe, and the Jurassic genera *Opthalmosaurus* and *Ichthyosaurus* represent the classic long-snouted, torpedo-bodied type. They moved through the water by moving their tail fins from side to side as their ancestors did, but they had more developed half-moon-shaped caudal fins akin to those seen in fast-moving fish like mako sharks, marlin, and tuna.

The reason ichthyosaurs evolved this way was determined by their ancestry. The way their forerunners moved on land was adapted for life in the water; what already existed was co-opted for new uses. The different modes of locomotion exhibited by the paddling plesiosaurs and mosasaurs illustrate other evolutionary alternatives, but the way all these reptiles swam was constrained by attributes of their ancestors. The same is true of cetaceans, and the reason for their distinctive method of swimming is just as fortuitous.

The ancestors of the early archaeocetes were terrestrial artiodactyls that carried their legs underneath their body. When they walked or ran their spine would have moved up and down, as in other quadrupedal mammals, and this simple fact of mammalian locomotion constrained the way the early ancestors of whales could have swum. When *Pakicetus* entered the water, with its long legs and relatively inflexible spine, it could not have swum like a fish or a lizard; it would have to have done the doggie paddle to get around.

Ambulocetus was a more proficient swimmer. Its expanded hands and feet were almost certainly webbed, and it probably swam by undulating its spine to a limited degree and paddling with its feet. For the earliest archaeocetes, each swimming stroke would have two parts, a power stroke that propelled them through the water and a recovery phase in which the limb was brought back into position for the next stroke.[55] This type of swimming does not allow for quick pursuit of prey, but by undulating its spine *Ambulocetus* could have kept moving even during the recovery phase of a stroke. Living otters provide a good model for this type of swimming, as they use both the sinuous undulations of their spines and their limbs to move through the water.

Ambulocetus was probably more reliant on paddling than movements of its spine, though, because of a feature inherited from its land-dwelling ancestors. In mammals, the area of the spine that is connected to the hip is called the sacrum, and it consists of vertebrae that have fused together to give strength to the hips. For archaeocetes beginning to undulate their spines to move through the water, the sacrum limited their spine's range of motion. Increasing flexibility in this part of the spine would have been an advantage. A greater range of up and down motion, and thus more powerful propulsion, could be achieved by unfusing the sacral vertebrae.

This vertebral adaptation is exemplified by *Rodhocetus*. The unfused vertebrae of its spine even functionally resemble those of the tail, indicating their role in locomotion. *Rodhocetus* still had its sacral vertebrae fused to its hips, however, and this means that archaeocetes began undulating their spines to move through the water before they stopped

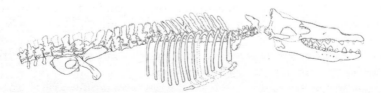

FIGURE 61 – The partially restored skeleton of *Rodhocetus* (the limbs were incomplete in the original description).

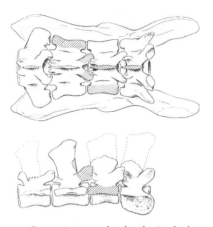

FIGURE 62 – The sacral vertebrae of *Rodhocetus*. Normally these vertebrae are fused in mammals, but in Rodhocetus they were unfused to allow greater flexibility of the spine.

using their limbs to propel themselves. Indeed, *Rodhocetus* and related genera like *Maiacetus* still had large, paddle-shaped feet, but another relative named *Georgiacetus* may reveal how early whales began using vertebral undulation more than paddling.

Georgiacetus had relatively large paddle-shaped rear limbs, but its spine was disconnected from its hips. Without the spine providing an anchor the hind limbs might not have been useful in swimming, but they could have worked as rudders to help direct the whale through the water. The fact that *Georgiacetus* and the bones of related archaeocetes have been found in the United States, thousands of miles from the epicenter of early whale evolution in Pakistan, indicates that it was a proficient enough swimmer to cross oceans, and it sat close to the transition from the protocetid-type whales to the oceangoing basilosaurids. By the time these whales evolved, the sacral vertebrae were entirely unfused, the hips were completely disconnected from the spine, and the hind limbs had become so reduced that they may have been functionless.[56] Ultimately cetaceans lost all external remnants of the hind limbs, probably because the heavy vestigial structures increased drag. Any such encumbrances would have been detrimental to predators that relied on speed to chase after their prey.[57]

The adaptation of early whales to the water also involved more subtle changes. Buoyancy is a major problem for terrestrial animals in the water; it is difficult to submerge or stay submerged with light bones filled with spongy tissue. Light bones require an animal to exert a lot of energy to stay under water, and this makes it very difficult to swim. In order for archaeocetes to fully adapt to aquatic life they would have to become neutrally buoyant, so that they would neither sink nor float when at rest in the water.

Some animals achieve neutral buoyancy as a result of swallowing stones (called gastroliths once they enter the stomach) that are the functional equivalent of ballast on a ship. Living crocodiles are among the animals that benefit from gastroliths in this way, but archaeocetes became neutrally buoyant due to changes in their bones. The cavities in limb and rib bones of terrestrial mammals are typically filled with marrow, but in many archaeocetes the cavities are filled with trabecular bone. Trabecular bone itself is not as dense as other types of bone, but this in-filling of bone cavities that otherwise would be empty made the bones as a whole denser and functioned as a kind of bone ballast.[58]

The ribs of *Indohyus* and *Pakicetus* show just this kind of bone restructuring. The development of bone ballast preceded other aquatic adaptations, but it had a trade-off. As the bones in the limbs and ribs of these animals became denser they would have become more brittle. Running on land with heavy, brittle bones would have been more energetically expensive and even risky, thus providing another reason for these animals to spend more time in the water. The amount of hypermineralization seen in the bones of *Pakicetus* suggests that it was doing just that.

The bones of later archaeocetes, like *Ambulocetus* and *Kutchicetus*, were even denser, and the heavy bone ballast in archaeocetes like *Rodhocetus* allowed them to devote more energy to swimming. Eventually, though, this trend would be reversed. *Georgiacetus* had a lighter bone density and marked the beginning of an osteoporotic trend, in which bone is thinner and lighter. This was carried even further in later marine whales like *Dorudon*, whose bone density more closely corresponded to that of living cetaceans. The reason for this reversal is attributable to life at sea. An extra heavy skeleton is a disadvantage during deep dives, where a large amount of energy is expended in search of food. The denser the skeleton a whale had, the more energy it would take to return to the surface before running out of breath. Making the skeleton lighter would allow whales to roam into new niches, not only feeding at the surface but also deeper down. Thus, natural selection favored a lightening of the skeleton.

The skeletons of extinct whales have filled in our understanding of how whales became adapted to life at sea, but not all the evidence comes from the fossil record. If living cetaceans are truly the descendants of terrestrial ancestors, then they should retain some vestiges of their ancestry. This is most certainly the case.

As with ichthyosaurs, the pectoral flippers of dolphins were adapted from arms and hands that were once used for terrestrial locomotion. Many cetaceans have additional bones in some of their fingers (primarily the second and third fingers) that lengthen their pectoral paddles, but the bones within their flippers are only modified versions of the front limbs of their terrestrial ancestors. Living whales retain vestiges of their hips and hind limbs, too. In cetaceans, the muscles of the tail important to propulsion connect to muscles in the abdomen, but the vestiges of the pelvis function as an anchor for muscles that attach to reproductive organs (an anatomical feature that is seen in other mammals, including artiodactyls). In this case the altered and reduced hips retain some of their original function, but some whales have exhibited archaic traits that are of no use whatsoever. In 1881, John Struthers published a paper called "On the Bones, Articulations, and Muscles of the Rudimentary Hind-Limb of the Greenland Right-Whale (*Balaena mysticetus*)," in which he discussed the vestigial hips, femur, and tibia found inside a right whale. The rudimentary leg bones were cartilaginous (which is why you don't usually see them in museum mounts) and appeared to have no function.

Why should these traits be preserved or even reappear if whales lost their external limbs long ago? While the morphology of whales has certainly changed, some of the genes that code for lost traits have been retained. This is because there is no single gene that could have been eliminated that controls the formation of legs. Instead, there are many genes involved in limb formation, some of which are involved in the development of other parts of the body. The multifunctionality of these genes helps to explain their endurance as well as why, for a short time, whales begin to grow legs in the womb.

At twenty-four days old, the embryos of spotted dolphins have both fore- and hind limb buds. By forty-eight days, however, the forelimb is well developed, with the beginnings of digits, but the hind limb bud is little more than a speck. It is resorbed into the body. What this suggests is that after whales started propelling themselves by oscillations of the tail and external hind limbs began to be selected against, there was a relatively sudden change in development that caused the cessation of growth and resorbtion of those limbs. Some of the genetic instructions for limb development remained, but they have been prevented from progressing far enough to create external hind limbs. An exception is a bottlenose dolphin with external pelvic fins described in 2006. It did not have fully developed hind limbs, but it illustrates how genetic instructions for "long lost" traits can be preserved and expressed due to changes in development.

In a similar fashion, baleen whales develop the beginnings of teeth only to have them resorbed during their early development. Even though adult baleen whales do not have any teeth, they still retain some of the genetic instructions for their formation. As we all learn in elementary biology, in DNA genes are made up of three-paired combinations of four letters, A, T, G, and C. Combinations of these letters have different functions, from coding for a protein to turning other genes on or off, but not all of the genome of any given organism is functional. If there is a mutation where a new letter is inserted into the code of a gene, it might create a stop codon, or a genetic stop sign that prevents that gene from being fully expressed. These stop codons might have different effects based upon where they occur along a gene, but when a gene loses its function it may pick up more and more mutations, becoming a "fossil gene" that slowly degrades away. In the case of living mysticetes, Thomas Deméré and colleagues found that they possessed two such fossil genes related to tooth production, called AMBN and ENAM. Both contained stop codons that prevented their expression. Even though mysticetes have not had teeth for millions of years, the genes for creating teeth are still there, slowly being eroded away by mutations.

The genetic information Deméré and his peers found formed a wonderful correspondence with the fossil record of baleen whales. Some of them, like the recently discovered *Janjucetus*, superficially resembled *Basilosaurus* more than a humpback whale. These frightening predators were the early relatives of today's filter-feeding giants, and the clues to this connection can be seen in the skull.

As odontocetes and mysticetes evolved, their skulls became longer, and certain bones became elongated in a trend called "telescoping." This is what caused the nasal openings of the whales to move from the front of the skull back over the eyes. If we were to line up a selected group of fossil whale skulls in the order in which the species appeared,

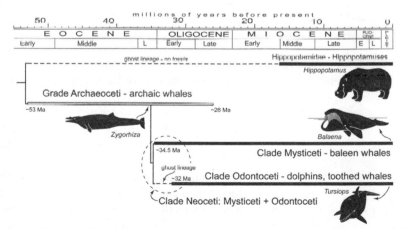

FIGURE 63 – A simplified evolutionary tree of whales over the past fifty-five million years. While archaeocetes persisted until about twenty-six million years ago, by that time the first representatives of modern whales (the odontocetes and mysticetes) had appeared.

from oldest to youngest, we would see that the early forms had a relatively short premaxilla and maxilla, the bones of the upper jaw that hold the teeth. Right behind the maxilla are the nasal bones, or the bones surrounding the nasal opening. Moving from the skulls of geologically older cetaceans to younger ones, the maxilla becomes longer and longer, pushing the nasal bones back on the skull until they are sandwiched together with other parts of the skull over the eyes.

The maxilla is important for another reason. In baleen whales the maxilla has extended so far back that part of it scoops downward and backward under the eye socket. This is not seen in odontocetes, but it is a characteristic feature of baleen whales. The skull of *Aetiocetus*, a baleen whale with teeth, shows just this condition where the maxilla extends backward to make up part of the margin of the eye. Stranger still, *Aetiocetus* probably had baleen.

Modern mysticete whales do not have teeth. Instead, they sift their food from the sea with a comb of hairlike structures that sprout out of plates attached to their jaws. In the past, however, the early relatives of modern baleen whales had teeth, and when it was first discovered *Aetiocetus* was an important confirmation that modern baleen whales evolved from toothed ancestors. What has only more recently come to

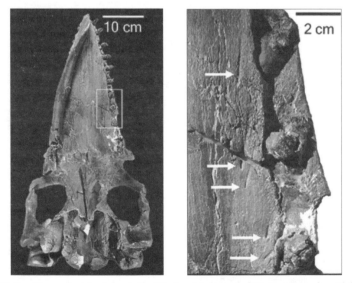

FIGURE 64 – The underside of the skull of the fossil baleen whale *Aetiocetus weltoni* (left) and a close-up of part of the jaw (right) showing foramina that would have nourished baleen with blood.

light, however, is that baleen evolved before the mysticetes fully lost their teeth. The hairlike, keratinous baleen of whales hangs from a set of plates that attach to the upper jaw, and these plates are nourished by blood vessels. The blood vessels come through a series of holes, or foramina, in the roof of the mouth, and *Aetiocetus* has the same telltale foramina seen only in modern baleen whales. How the baleen helped *Aetiocetus* feed is still unknown, (perhaps it was useful in filter feeding, or straining prey from mud stirred up from the sea bottom), but the unexpected discovery of an ancient whale with both teeth and baleen is a stunning confirmation of evolution.

There are some traits of modern cetaceans that were not direct adaptations to life in the sea, like the increases in brain size seen in dolphins thirty-four and fifteen million years ago (most likely a result of changes in the social lives of dolphins), but the general whale body form can be understood as an extreme modification of an ancestral hoofed mammal type to aquatic life. This change happened extremely fast. Only ten million years separate the wolflike *Pakicetus* and the fully aquatic *Basilosaurus*. Fortunately the fossil record has preserved an extraordinary diversity of fossil whales. A confluence of fossil and genetic data has allowed us to understand a transition that was once mysterious, but there are still things to learn about lineages that were thought to be well known.

Behemoth

"Behold now behemoth, which I made with thee; he eateth grass as an ox. Lo now, his strength is in his loins, and his force is in the navel of his belly. He moveth his tail like a cedar: the sinews of his stones are wrapped together. His bones are as strong pieces of brass; his bones are like bars of iron. He is the chief of the ways of God: he that made him can make his sword to approach unto him."
— JOB 40:15-19 (King James Version)

"It appears somewhat extraordinary, at the first view, that we should discover manifest proofs of there having existed animals of which we can form no adequate idea, and which in size must have far exceeded any thing now known upon the earth; and those too, in climates where elephant (the largest animal now in existence) is never found."
— GILBERT IMLAY, *Topographical Descriptions of the Western Territory*, 1792

The war was finally over. On September 3, 1783, representatives of the American Congress of the Confederation and a delegation representing King George III of England signed the Treaty of Paris, formally bringing an end to the war for American independence that had begun in 1775. It would take more than seven months until both parties had ratified the agreement, but by then the cannonade and musket fire that shook the nascent American nation had long ceased. With the war for freedom over, the United States of America now had to fight for the respect of the European powers that had allied with it and fought against it during the previous decade.

Careful political maneuvering was pivotal to fostering the growth of the United States, but science would also have a role to play. This would bring one of America's Founding Fathers, the polymath Thomas Jefferson, into direct conflict with one of the greatest naturalists of France, Georges-Louis Leclerc, Comte de Buffon. In this intellectual battle, in which natural history had everything to do with national pride, facts from zoology and paleontology would be essential elements in the arsenals of both parties.

Buffon had fired the opening shot even before the onset of the American Revolution. The research required to write his great, multi-volume series *Historie naturelle* had acquainted Buffon with the varied natural productions of the world, but when he looked from the Old World to the New he noticed a great disparity. The New World had none of the impressive animals that had for so long captivated the European imagination. In the ninth volume of the series, published in 1761, Buffon wrote of the Americas,

> In general, all the animals there are smaller than those of the old world, & there is not any animal in America that can be compared to the elephant, the rhinoceros, the hippopotamus, the dromedary [camel], the giraffe, the buffalo, the lion, the tiger, etc.

Not only that, but in Buffon's estimate the Americas were not even capable of supporting such prodigious beasts. The New World was a hemisphere marked by degeneracy, proved by the fate of the robust domestic animals European settlers had introduced to North America. They did not flourish, and Buffon suspected that an inhospitable climate was to blame for the corruption of the European breeds. Buffon pointed out that the most magnificent animals were found in hot, equatorial regions. Clearly the warm temperatures better stimulated the production of life, and Buffon reinforced his hypothesis with evidence from prehistory. Despite its present relatively cool climate, the petrified bones of tropical animals such as elephants, hippos, big cats, and rhinos had been discovered in western Europe. (The remains of the elephants were sometimes argued to have belonged to war elephants brought into Europe by conquering Romans during antiquity, but the same explanations could not be applied to the other animals.) Apparently they had flourished during a past time when Europe was much warmer.

Buffon explained this climate change in his 1778 tome, *Epoques de la nature*. After conducting experiments with globes made of various metals that were heated white hot and then left to cool, Buffon proposed that the world was formed in a molten state and had been slowly cooling ever since.[59] The fossil hippos, lions, rhinos, and other tropical beasts had inhabited Europe during a time when the world was still relatively warm but their ranges had subsequently shrunk as the earth continued to cool. That the New World did not harbor animals of equal stature meant that it was unlikely that they were present during the past; it had always been a brutish, infertile place devoid of the great natural wonders of the Old World. And if this was the state of the nat-

FIGURE 65 – Thomas Jefferson painted in 1800 by artist Rembrant Peale.

ural world, what would become of the colonists who chose America as their home? Would they, too, degenerate into an inferior society?

Thomas Jefferson could not abide this insult. It mattered little that, after conversing with Benjamin Franklin on science, Buffon changed his views on American degeneracy. The New World, and North America in particular, was often denigrated as being inferior to Europe. Jefferson was driven to prove that America was as glorious, as wonderful a land as any other.

Jefferson documented the natural richness of America, along with his views on politics and society, in *Notes on the State of Virginia*, a book that had its roots in a list of questions that had been given to representatives of each of the thirteen colonies by the secretary of the French delegation to the United States. The responsibility for responding on behalf of Virginia eventually found its way to Jefferson. Published in France in 1784, the book was his minutely detailed response to the French government and Buffon alike. Among the various descriptive passages was data on the size of North American animals like bear and deer, and though some of the reports were inflated (like domestic cows weighing over a ton), Jefferson provided ample evidence that America was just as vigorous and stimulating as anywhere else in the world.

Jefferson sent a copy of *Notes* and the skin of an exceptionally large cougar to Buffon. This was enough to gain him an invitation to dine with the French naturalist at the city's botanical garden, the Jardin du Roi. In conversation Jefferson found Buffon almost entirely unfamiliar with the large animals of America, and Jefferson soon wrote back to General John Sullivan, the governor of New Hampshire, asking that the skin and bones of the largest moose that could be found be sent directly to him in Paris.[60]

But the moose was not the biggest animal that Jefferson herded into supporting his case for American robustness. Though no one had seen one alive, there were rumors that an even larger animal still roamed the largely uncharted expanses of the American interior. The only sign that it had ever existed were enormous bones, many of which had been found at what would come to be known as "Big Bone Lick," in Jefferson's home state of Virginia.[61]

The graveyard of the enormous beasts in the Virginia backcountry had come to the attention of Europeans several decades before. In 1735, a small group of French-Canadian soldiers were marching from Quebec toward New Orleans with their Native American allies from the Abenaki tribe (who actually made up the bulk of the fighting force of over 400 men). Under the command of Charles le Moyne, their mission was to drive back the Chickasaws, a tribe allied with the English, who were threatening New Orleans and had made passage toward the port city along the Mississippi all but impossible.

By summer the force was heading southwest along what is now known as the Ohio River. It was too long a trip to be covered by packed provisions, so Abenakis were often dispatched from the group to bring back game. One night, at dusk, as the army was camped along the river not far beyond where the city of Cincinnati, Ohio, would one day be founded, the hunters brought back not only meat, but several enormous teeth, a femur nearly as tall as they were, and dark-colored tusks. The bones were not entirely unfamiliar to the Abenakis. Their legends told of ancient monsters vanquished by heroes, enormous elk, and water monsters. The French-Canadian soldiers did not believe these stories, but recognized that the bones might be valuable and carried them along.

Eventually the army arrived at their destination on the rim of the Gulf of Mexico, and from there in 1740 Charles le Moyne brought the fossil bones with him to France. When the French anatomist Louis Daubenton presented his conclusions about the fossils to the French Royal Academy in 1762, he asserted that the bones belonged to at least two different animals. Clearly the tusk and the femur belonged to some kind of elephant, but the molars, similar to the molars of hippos from Africa, seemed better suited to pulverizing flesh and bone than grinding grass.[62] This would give any would-be explorers of North America's interior cause for alarm, since both the elephant and the meat-eating hippo probably still lurked somewhere in the American wilderness.

That this type of molar tooth and tusks had often been found together led Jefferson to doubt Daubenton's assessment, but he did agree that the species was still alive somewhere west of the Mississippi. "Such is the economy of nature," Jefferson wrote, "that

FIGURE 66 – The molar tooth of an American mastodon, originally known as the "American Incognitum." The rough bumps and ridges of the teeth led some eighteenth and nineteenth century naturalists to believe that the mastodon was a carnivore.

no instance can be produced of her having permitted any one race of her animals to become extinct; or her having formed any link in her great work so weak as to be broken."

Jefferson could also appeal to the fact that many Native American tribes, including the Abenakis, passed down legends about monstrous creatures bigger than the biggest bison. The only conclusion could be that Native Americans had seen them during recent history, though where the creatures had gone was unknown. Even so, if the "American *Incognitum*" was still alive then it certainly rivaled or surpassed even the largest animals of the Old World. Jefferson could not provide a living specimen or its pelt, but he could convincingly use it to deflect any remaining remarks about American "degeneracy."

Jefferson was not only interested in the animal for political reasons. He was an avid naturalist and was fascinated by fossils. That the *Incognitum* might still live only added to the mystery of the Louisiana Territory, which Jefferson acquired for the United States during his first presidential term in 1803. When Jefferson sent Meriwether Lewis and William Clark to explore the purchased land, among other things he told them to keep an eye out for the lumbering *Incognitum* as well as any interesting fossils that he could add to his collection in what would become the East Room of the White House.

If Lewis and Clark did bump into the elephantine beast, naturalists believed, they would be lucky to escape with their lives. Even after it became generally agreed that the molars and tusks of the kind Moyne found belonged to one animal, Daubenton's interpretation that the molars indicated an animal of carnivorous habits stuck. This interpretation was most colorfully argued by the naturalist George Turner during a 1799 meeting of the American Philosophical Society in Philadelphia. Said Turner of the carnivorous elephant:

> May it not be inferred, too, that as the largest and swiftest quadrupeds were appointed for his food, he necessarily was endowed with great strength and activity? That, as the immense volume of the creature would unfit him for coursing after his prey through thickets and woods, Nature had furnished him with the power of taking it by a mighty leap? – That this power of springing a great distance was requisite to the more effectual concealment of his bulky volume while lying in wait for his prey?

The image of a rapacious, bone-crushing elephant leaping out of the woods onto deer, bison, and even humans stretched credulity, but such were the ways of nature.[63] If the *Incognitum* had teeth well suited

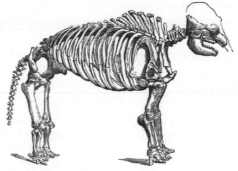

FIGURE 67 – The nearly complete skeleton of an American mastodon, as figured in Cuvier's description of fossil elephants.

to shearing flesh, then the rest of its body and habits would have been fine-tuned to make it a successful hunter. The existence of such a creature was not cruel but a sign of the beneficence of God or Nature. Such predators weeded out the old and the sick, and great appetites could not exist without the means to satisty them.

Turner's vision of the *Incognitum* was not to last. As the polymath Benjamin Franklin commented when he saw the hefty molars of the beast in 1768, such teeth "might be as useful to grind the small branches of Trees, as to chaw Flesh." Debate on this point went back and forth for years, but it was eventually determined that the creature was an herbivore and not a bloodthirsty elephant. It crushed tree branches in its jaws rather than limb bones. Even so, this did not change the interpretation that it was a creature perfectly molded by nature that would not be allowed to disappear. It would be Georges Cuvier who would destabilize the "economy of nature" that other naturalists of the time fervently believed in.

By the close of the eighteenth century several ill-defined species of elephant were recognized. There were the living African and Asian varieties, the American *Incognitum*, and another type of fossil elephant often found in the permafrost of Siberia. This latter animal, like its American counterpart, was surrounded by myth and legend, but the bones were often thought to have belonged to one of the living types that had wandered further north when the world was still warm.

In 1796, Cuvier set about determining, once and for all, whether these types separated by time and space were really just one species or several. He presented his results in what would become a landmark paper, "Memoirs on the Species of Elephants, Both Living and Fossil." Incredibly, Cuvier concluded that each type was a distinct species. Through comparative anatomy he was able to determine that the African and Asian elephants were distinct from each other, and the fossil types differed just as strongly. The North American creature, identified by the breastlike bumps of its teeth, would later be named *Mastodon* by Cuvier.

The creature from Siberia, on the other hand, would be called the *mammoth*, and Cuvier's identification would be supported three years later when the Russian botanist Mikhail Ivanovich Adams discovered the skeleton of a mammoth—with tatters of flesh still clinging to it—near the mouth of the Lena River in Siberia. While the mammoth showed some resemblance to the Asian elephant, Adams's specimen revealed that the living animal had been covered in a coarse, shaggy coat that seemed better adapted to the modern frozen tundra than the past global hothouse Buffon had proposed. What puzzled naturalists, however, is where these elephants had come from and what had become of them.

The bones of the mastodon and the mammoth were both found in places elephants did not inhabit in Cuvier's time, and none of the traditional explanations for this disparity held up to scrutiny. No staunch biblical literalist would like to admit that Noah had decided to leave some species behind, and the bones were too different from those of living species to make credible Buffon's proposal that they were the remains of living species that had once inhabited different parts of the world during warmer times. Likewise, it was difficult to take seriously Jefferson's belief that such animals might still be living in unexplored parts of the globe. In an age of expansion, exploration, and empire there were few places left in the world for such large animals to hide. Extinction was the only reasonable explanation for the pattern Cuvier saw:

> All these facts, consistent among themselves, and not opposed by any report, seem to me to prove the existence of a world previous to ours, destroyed by some kind of catastrophe. But what was this primitive earth? What was this nature that was not subject to man's dominion? And what revolution was able to wipe it out, to the point of leaving no trace of it except some half-decomposed bones?

This was a bold statement for the young anatomist to make. Even though the concept that there might be "lost species" had been toyed with previously, Cuvier had finally presented evidence for it.

While Cuvier's belief that there was nowhere left for previously unknown large animals to hide was premature, those who held out hope that prehistoric species might still be found had their faith eroded by the discovery of even stranger fossil creatures. Among other distinct fossil species Cuvier also described the crocodilelike "Monster of Maastricht" he called *Mosasaurus* and an enormous sloth from Buenos Aires, Argentina, he christened *Megatherium*.[64]

Even Jefferson reconsidered the disappearance of species. In an

1823 letter to his friend and rival, John Adams, Jefferson couched extinction within his belief of an orderly, regulated natural universe.

> It is impossible, I say, for the human mind not to believe that there is, in all this, design, cause and effect, up to an ultimate cause, a fabricator of all things from matter and motion, their preserver and regulator while permitted to exist in their present forms, and their regenerator into new and other forms. We see, too, evident proofs of the necessity of a superintending power to maintain the Universe in its course and order. Stars, well known, have disappeared, new ones have come into view, comets, in their incalculable courses, may run foul of suns and planets and require renovation under other laws; certain races of animals are become extinct; and, were there no restoring power, all existences might extinguish successively, one by one, until all should be reduced to shapeless chaos.

Though beautiful, the universe was not perfect, and in Jefferson's view it required some preserving force to fix what had become damaged and replace what had been lost. Whatever this force was, whether it was deity or natural law, it acted as a tinkerer, never putting things back exactly as they were.

With the science of paleontology established, European naturalists began to search their own backyards for more fossils. What they found was that their countries were founded upon vast accumulations of elephant bones. Where living and fossil types had been lumped into one species, there was now an explosion of proposed extinct types, such as a predecessor to the mammoth, *Elephas meridionalis*, and another widespread form, *Elephas priscus*. Cuvier and his English counterpart Richard Owen doubted the validity of some of these new species, but even so, the abundance of extinct elephants in Europe was remarkable.

Even more fossil elephants were found in distant territories controlled by European powers. In 1830, the young botanist and paleontologist Hugh Falconer left England to act as an assistant surgeon for the British East Asia Company in India. While most of his time was occupied by his duties as a doctor, or in the botanical gardens at Suharunpoor, when he had a few moments to spare he picked up the fossils that littered the nearby Siwalik Hills. Many of the deposits Falconer traversed were Cenozoic in age, or represented the Age of Mammals which succeeded the previous age in which reptiles swarmed over the globe. Falconer was hardly the first to find fossils here. Local legends held that the bones of giants were embedded in the hills, and

like the bones found by the Abenakis along the Ohio River these remains had come from numerous prehistoric elephants. Falconer wrote:

> What a glorious privilege it would be, could we live back—if only for an in-
> stant—into those ancient times when the extinct animals peopled the earth!
> To see them all congregated together in one grand natural menagerie—these
> Mastodons and Elephants, so numerous in species, toiling their ponderous
> forms trumpeting their march in countless herds through the swamps and
> reedy forests!

Falconer described these extinct behemoths in more detail in his
uncompleted monograph series *Fauna Antiqua Sivalensis*. Following in
Cuvier's footsteps, Falconer's work was more descriptive than theoretical, but he identified some interesting patterns among the fossils he
had found. The most immediately apparent was that the elephants of
the past were far more diverse than living forms. Modern elephants
were only the "ragged remnant representation of the rich garment of life
with which the continent was formerly clothed."

Extinction had swallowed all of these fossil species, but the force
that brought them into existence was a mystery. Falconer rejected the
popular idea that complete faunas sprung forth instantaneously in the
wake of catastrophes. It was an insult to the Creator to think that the
world required such constant revisions. The continuity of the fossil
record suggested something different.

By the 1850s the remains of prehistoric elephants had been dug up
from locations all over the northern hemisphere for almost the entire
span of the Cenozoic (at least as it was then understood). There was the
American mastodon, an extremely large variety from Europe with
recurved tusks sticking out of its lower jaw named *Deinotherium*, and
the numerous species Falconer had dug up in the Siwalik Hills, but the

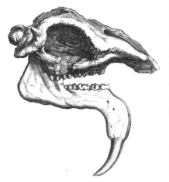

most widely distributed type was the
mammoth. In a study published in 1845,
the French geologist Étienne d'Archiac
proposed that mammoths ranged from
western Europe through Asia, over the
Bering Strait into western Canada, and
down into the southwestern United
States. While they lived during the latter

FIGURE 68 – The skull of *Deinotherium*, one of the first
fossil proboscideans ever discovered.

parts of the Cenozoic, divisions of time known as the Pliocene and Pleistocene, as a single species they would have persisted over a huge expanse of time.

Falconer suspected that there was something amiss with this pattern, and in his reexamination of fossil elephants he found that other naturalists had perhaps been a bit overzealous in lumping the various fossils into just one species. Using delicate dental characteristics as his guide, Falconer was able to distinguish several species in Europe alone. There was not just one species of mammoth but several closely related types from different places and times.

What was even more startling, however, was that some of the species Falconer identified seemed to represent intermediate forms. The grinding teeth of the different species could be organized into a linear progression from the rough, bumpy teeth of the mastodon to the flat, crenulated molars of living elephants. Not all the fossil teeth fit comfortably within this arrangement, but the transitional features Falconer identified still created a continuous series of forms.

Falconer had created what would appear to be an evolutionary series, yet he doubted that any of Europe's fossil elephants could have been ancestral to later types. His reasoning was that none of the elephant species he had studied showed any variations that would reveal them "approaching," or smoothly grading into, another species in form. Every species was an island that remained unchanged for long periods of time. This, in Falconer's mind, ruled out evolution by natural selection as proposed by Charles Darwin. But Falconer did not reject evolution altogether; he eschewed special creation and believed that mammoths certainly were "the modified descendants of earlier progenitors." The only question was what, if not natural selection, had precipitated this change.

Many other paleontologists were in accord with Falconer. Even though a variety of fossil elephants were known by the end of the nineteenth century, the early evolution of the Proboscidea, the group to which elephants and their extinct relatives belonged, was entirely unknown.

Due to the contingent facts of history, however, the leading 19th century paleontologists had primarily worked and studied near areas that preserved relatively recent fossil elephants. The sought-after early ancestral types would be found elsewhere. During the late nineteenth century, England and France gained increasing control of the Egyptian government, though they were resented by the Egyptian people. In 1882 this tension escalated and erupted into the Battle of Tel el-Kebir, in which British soldiers were dispatched to protect the crown's interests. The British overran the Egyptian soldiers, allowing England to take control of the country and opening the land to visiting geologists.

At the turn of the twentieth century, the chronically ill British paleontologist Charles Andrews, seeking a warm climate on doctors' orders, set out to search for fossils in the de facto colony. Just a few years before, Andrews's colleague Hugh Beadnell had found numerous fossils in the Fayum desert, about eighty-one miles south of Cairo, and with Beadnell's help Andrews was to find many more.

Most of the fossils of the Fayum were from much older parts of the Cenozoic than were preserved back in Europe or in Hugh Falconer's old stomping grounds in India. The deposits were of Eocene and Oligocene ages with many fossils spanning the boundary between the two at about thirty-seven to twenty-eight million years old. In contrast to the modern, arid conditions, though, during this time the Fayum had been home to tropical forests and swamps, and in the vicinity of those swamps lived some of the earliest proboscideans.

Andrews and Beadnell found several early elephant relatives. There was *Barytherium*, an animal the size of an Asian elephant that had two sets of four very small tusks oriented upright in its jaws. This was an arrangement similar to that of *Moeritherium*, a smaller Fayum proboscidean that had an enlarged set of incisor teeth but a build like a hippo. Their neighbors, *Palaeomastodon* and *Phiomia*, were decidedly more elephantlike. These forms had longer upper and lower tusks, and

FIGURE 69 – The skull of *Palaeomastodon*, with a restoration of its head by Charles R. Knight.

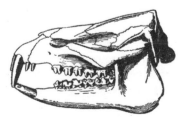

FIGURE 70 – The skull of *Moeritherium*, one of the earliest proboscideans known from the Fayum desert.

situated further back on the skull a nasal opening that undoubtedly supported a rudimentary trunk.

While there was some disagreement about the exact placement of these creatures in relation to other fossil elephants, they seemed to solve the mystery of the group's early evolution. Proboscideans had originated with stout, hippolike forms such as *Moeritherium* before evolving into the much more elephantlike *Palaeomastodon*. From there elephant evolution exploded, with a profusion of four-tusked creatures such as *Gomphotherium* giving rise to mastodons, mammoths, and eventually modern elephants. Thus the great march of elephant evolution seemed all but complete. In a 1914 summary of elephant evolution the American paleontologist Erwin Hinckly Barbour stated,

> The genealogy of [elephants] is now so well known to naturalists, that it is interesting to note in the writings of Cope and others of twenty-five years ago, that the intermediate proboscideans are entirely lost, and the phylogeny of the order absolutely unknown. As a reward of zeal, the genetic gaps are being filled so rapidly, that ultimately knowledge of the history of the Proboscidea must be as well known as that of the Equidae [horses].

But not all the known fossil proboscideans fell into this straight line of elephant development. *Barytherium*, known from only a handful of bones at the time, did not fit the pattern that was expected and was pushed off to the side. Likewise *Deinotherium*, the large "anchor-tusked" elephant relative discovered in the early part of the nineteenth century, was also seen as an evolutionary dead end only distantly related to living elephants.

Then there were the "shovel-tuskers." During the 1920s Barbour described *Amebelodon*, a proboscidean with a long, scooplike lower jaw that lived about nine to six million years ago during the Miocene. Even more widely dispersed was a similar form that lived around the same time, *Platybelodon*, that also had lower jaws and teeth shaped like a shovel. At the time naturalists took this to mean that both *Amebelodon* and *Platybelodon* lived in wetlands, where they used their jaws to scoop up soft water plants, but they were so strange that clearly they could not have been ancestral to any of the mammoths or ancestors of living elephants.[65]

This led paleontologists to use two competing sets of imagery when mapping the ancestry of elephants. The public was presented with an image of straight-line elephant evolution from smaller, generalized ancestors to larger, specialized descendents. The truth of evolution could

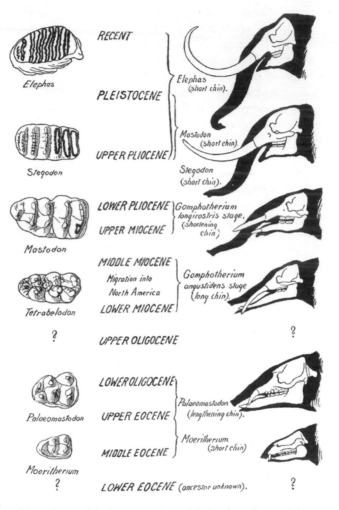

FIGURE 71 – The evolution of elephants as envisioned during the early twentieth century.

not be denied in the face of such illustrations. Yet these images were often accompanied by wildly branching bushes of elephant diversity that housed all the disparate lineages of the trunk-bearing herbivores.

These seemingly competing forms of imagery visually complemented how many paleontologists viewed evolution during the early twentieth century. Rather than being affected by natural selection, evolution was believed to be driven by internal forces toward particular endpoints. There was a "March of Progress" from the primitive to the advanced, words that were just substitutes for "lower" and "higher," which was thought to be the main stem of evolution. Any creature that could not be slotted into this trajectory was considered to be a dead-end offshoot, and these creatures typically received comparatively little attention.

A perceived side effect of the internal perfecting principle, however, was that many lineages made attempts at reaching the same goals. If they failed, like *Deinotherium* and *Platybelodon*, it was because they had somehow been knocked off the progressive path evolution was proceeding along. Any traits they shared with similar creatures could be explained away under the aegis of "parallel development."

This view was most explicitly outlined by the American paleontologist H. F. Osborn, especially in his massive, posthumously published 1936 monograph on fossil elephants simply titled *Proboscidea*. As envisioned by Osborn, fossil elephants could be arranged in a palm-frond pattern. From one central point there was an explosion of elephant types, each proceeding in a straight line from primitive to advanced forms during the course of time. There were a few branching points here and there, but the overall picture was one of a diverse collection of elephants all struggling upward toward a particular endpoint.

Even though Osborn had been a powerful force in paleontology in the early twentieth century, however, by the time of his death in 1935 his evolutionary views were outmoded, even among paleontologists who had worked under him at the American Museum of Natural History, such as W. K. Gregory and W. D. Matthew. The new evolutionary synthesis grounded in evolution by natural selection was emerging.

The task of reinterpreting Osborn's evolutionary scheme would be undertaken by G. G. Simpson, one of the chief architects of the new evolutionary synthesis, in his comprehensive 1945 monograph *On the Principles of Classification and the Classification of Mammals*. In the case of proboscideans, Simpson identified that there was an early radiation of bizarre forms, and while most of them became extinct twenty-three million years ago, at least one branch formed the base for the array of mastodons, shovel-tuskers, mammoths, and ancestors of living elephants. Elephant lineages were not competing with each other to reach an end goal, but a series of evolutionary divergences within a nested hierarchy that could be traced back into the Eocene. Proboscideans were a widespread, prolific group that flourished until very recently, and echoing the statements Falconer made a century before, Simpson wrote:

Although this order has living representatives, familiar to all, the two surviving genera are relicts of a dying group. . . . The order had a much greater
role in faunal history than one would dream from this impoverished representation, and it formerly occurred in bewildering numbers and variety
over the whole of all the continents, except Australia, and on a number of
islands.

The Proboscidea had not sprung from nothing, however. They were
mammals that would have to be anchored within the mammalian family tree. At first glance elephants seem to be different from almost every
other kind of mammal. Their long-ago grouping with other thick-skinned
mammals, such as rhinos and hippos, as "pachyderms" crumbled when
viewed from an evolutionary perspective. With the development of comparative anatomy, however, naturalists began to notice similarities between living elephants and another group of odd mammals, the sirenians,
or the aquatic mammals popularly known as manatees and dugongs.
Males of both manatees and elephants have internal testes ("the sinews
of his stones are wrapped together," per the description in Job), for
example, and females of both groups develop breasts that are closer to
their chests than their abdomen. And while elephants took the feature
to great extremes, some sirenians also had enlarged, tusklike second
incisors. Along with a few other features, such as eyes oriented more
forward on the head, these specialized characteristics solidified the
relationship between elephants and sirenians. Comparisons of elephant
and sirenian DNA show that they are more closely related to each other
than any other type of mammal.

But, as recognized by Falconer and Simpson, the organisms inhab-
iting the world today are remainders of past radiations, many branches
of which have become extinct. Manatees and elephants are each other's
closest living relatives, but both are also closely allied to two groups
that have long been extinct: the embrithopods and the desmostylians.
The most famous representative of the former, among the numerous
fossils Andrews and Beadnell found in the Fayum desert, was an early
Oligocene mammal with two immense facial horns called *Arsinoitherium*. The latter were marine mammals that lived from about
thirty million years ago until about seven million years ago. Found primarily at sites bordering the modern Pacific Ocean, these mammals had
heads like hippos but bodies more similar to those of sea lions. As with
Arsinoitherium and its close relatives, their exact relationships to elephants and manatees are unknown, but the desmostylians do share specialized features that associate them more closely with those groups

than with other mammals. Proboscideans, sirenians, desmostylians, and embrithopods can all be placed into a group called the Tethytheria, and each of these groups evolved and dispersed around the edge of the ancient Tethys sea.

Frustratingly, however, the earliest representatives of each group have been elusive. *Moeritherium* and *Palaeomastodon*, creatures thought to represent the beginning of proboscidean evolution, were too specialized to be the earliest elephant ancestors. The first animals that could have rightly been called proboscideans would have been small, lacked large tusks, and had no trunk to speak of. In 2009 paleontologist E. Gheerbrant described what is probably the oldest proboscidean yet found. The sixty-million-year-old remains from Morocco came from a rabbit-sized animal Gheerbrant named *Eritherium*. It possessed key proboscidean characteristics, such as slightly enlarged second incisors, but there is no sign that it had a trunk. At present it is impossible to know whether it was ancestral to any later species, but given its age it probably represents the form of some of the early proboscideans that evolved in the wake of the extinction of the dinosaurs.

It was likely a creature like *Eritherium* that gave rise to the diversity of elephants found in the Fayum desert. Reflected in Simpson's classification, the relatively early radiation of proboscideans covered a wide range in body size and possessed different arrangements of tusks and trunks. What is especially curious, though, is that some of these early forms appear to have spent a good deal of time in the water.

Ever since the time of its discovery, the vaguely hippo-shaped *Moeritherium* has been thought of as a swamp-dwelling creature. Dead animals can be washed into bodies of water, however, and many fossil animals that were once thought to be aquatic have been shown in later studies to prefer life on land or in the trees. In the case of *Moeritherium* and its relative *Barytherium*, though, the best evidence that they spent much of

FIGURE 72 – The partial skull of *Eritherium* as seen from the front (top) and underneath (front of the skull facing up).

their time in the water comes not from their overall body shape but their teeth.

In 2007 A. G. Liu, Erik Seiffert, and Elwyn Simons looked at the details of carbon and oxygen isotopes retained within the teeth of *Moeritherium* and *Barytherium* found in thirty-seven-million-year-old deposits in the Fayum desert. The amount and type of these isotopes were influenced by what kind of food the animals ate (carbon) and even whether they spent a good deal of time in the water (oxygen). (Similar techniques have been used to determine the pattern in which early whales moved further from shore and whether prehistoric horses were grazers or browsers.) This is possible because enamel is often less altered during fossilization than bone, thus preserving a record of the chemical components inside the teeth.

What the team found was that the carbon and oxygen isotope values of *Moetherium* and *Barytherium* were more similar to each other than any other mammals found from the same deposits. The oxygen isotope values in each creature, especially, showed little variation, a pattern seen in aquatic or semi-aquatic animals where much of the isotopic oxygen comes in through the animal's skin from the surrounding water. Terrestrial animals, which are not typically submerged, have more variable oxygen isotope values as the isotopic oxygen in their bodies comes from a wider variety of sources. While the authors were cautious in their assessment, stating that some factors (such as the amount of water an individual animal drank) could influence the isotope values, it appears that both genera were semi-aquatic and frequently ate water plants. *Barytherium* stayed closer to shore, chewing on the tough plants that grew at the water's edge, while the isotope values from *Moeritherium* suggest that it spent more time in the water and fed on more succulent plants that could only be reached by wading farther out.

While many of the early proboscidean lineages became extinct, one gave rise to more elephantlike forms such as *Palaeomastodon* and *Phiomia*. These and all their later relatives belonged to a group called the elephantiformes, marked by the possession of a nasal opening high up on the skull for the attachment of a trunk; enlarged, forward-facing tusks; and the loss of a few teeth, meaning that their premolars and molars carried the full duties of chewing.

This masticatory transition can be seen in another recently described fossil from eastern Africa named *Eritreum*. At twenty-seven million years old, *Eritreum* was a relatively small, four-tusked creature that shared traits with both earlier elephantiformes like *Palaeo-*

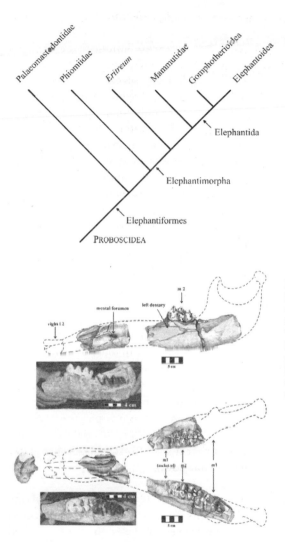

FIGURE 73 – A simplified elephant family tree (highlighting the relationship of *Eritreum* to other proboscideans). Living elephants are contained within the Elephantoidea, the diverse gomphotheres within the Gomphotheroidea, and mammoths within the Mammutidae.

FIGURE 74 – The restored lower jaw of *Eritreum* (with photos of the actual specimens) as seen from the side and from above. The third molar had only just begun to erupt and, as seen in the top photo, the scientists had to dig into the lower jaw to fully expose it. Artwork by Gary H. Marchant.

mastodon and the later group containing mastodons, mammoths, and living elephants called elephantimorphs.[67] In its lower jaw it only possessed a pair of tusks (enlarged second incisors) and a set of molars, yet a tooth still embedded in the lower jaw of the first described specimen shows that it had a trait similar to what is seen in modern elephants.

African and Asian elephants are among the longest-lived of all mammals. It takes a lot of time and energy to reach such prodigious size, and this requires that elephants consume huge quantities of relatively low-quality foods such as leaves and grass. Their teeth are worn down severely by all that chewing and do not regenerate.[68]

But elephants have an adaptation that allows them to avoid wearing down their teeth too fast. Rather than popping out of the gums all at once, their teeth erupt sequentially with only one grinding molar in use at a time. As subsequent teeth develop in the jaw they eventually push out the previous, worn-down tooth, ensuring that elephants have fresh teeth for much of their lives (though some exceptionally old individuals have been observed to grind down their very last set).

An early incarnation of this adaptation can be seen in *Eritreum*. While its molars were intermediate in size between proboscideans like *Palaeomastodon* and the first true mastodons, it did not get all of its adult teeth at once. In the specimen examined by scientists, the third molar was still developing inside the jaw but there was no room for it in the lower jaw. As it emerged, it would have pushed the second molar forward, and the first molar, by then worn down considerably, would have been pushed out of the jaw. As far as is presently known this adaptation was not present in early elephantiformes, so *Eritreum* appears to have been very close to the common ancestor of the profusion of mastodons that would spread all over the world.

While the name "mastodon" was first applied to what was once known as the American *Incognitum* (or *Mammut americanum* to today's paleontologists), today the term is used for a diverse array of elephants with coarse ridges on their molars. Despite this shared tooth form, however, mastodons came in a variety of shapes and sizes. Some, such as *Gomphotherium*, had low skulls with two upper and two lower tusks that stretched forward to give it a long profile. The shovel-tuskers *Platybelodon* and *Amebelodon* were a modification of this form, as were later, short-faced mastodons such as *Stegomastodon*.

With the exception of the long-lived genus *Deinotherium*, it was primarily the mastodons that left Africa during the Miocene and began to disperse through Eurasia to North and South America. They were so prolific, in fact, that in many places several species and even genera lived in the same habitats at the same time, each feeding on different types of plants. Some, like the wide-ranging *Gomphotherium*, were mixed feeders while other proboscideans that shared the same landscape were more exclusively browsers or grazers.

Proboscideans remained widespread through the following Pliocene and Pleistocene epochs, many of them being adapted to a grazing diet as the world cooled and grasslands spread across the globe. New forms

FIGURE 75 – A restoration of *Eritreum* (foreground) with the much larger *Gomphotherium* (background).

continued to evolve alongside more archaic lineages, including the one containing the American mastodon. Though it survived until about 10,000 years ago, the American mastodon was a remnant of a lineage that had split off many millions of years earlier. By the Pliocene, however, some parts of the earlier mastodon radiation had been lost, and a new diversification event in Africa would give rise to the most famous representatives of near prehistory, the mammoths.

Though the genus *Mammuthus* first evolved in Africa around four million years ago, much of mammoth evolution took place in Eurasia and, much later, North America. The dispersion of mammoths started when a mammoth adapted to warm climates, *Mammuthus rumanus*, moved out of Africa around three million years ago. It spread out through Eurasia (perhaps as far as China). One population of this species might have been ancestral to the species *M. meridionalis*.

M. meridionalis has long been considered the progenitor of the much larger steppe mammoth, *M. trogontherii*, which is thought to be ancestral to both the smaller woolly mammoth (*M. primigenius*) and the Columbian mammoth in North America (*M. columbi*). For a time this evolutionary pattern seemed quite simple, with the mammoths developing shorter, high-domed heads and an increased number of folds to their grinding molars over time, but recent reexaminations have shown that mammoth evolution could not be so easily fitted into a single-file pattern.

M. meridionalis was a long-lived and widespread species, but some of the later members of this species look very similar to early steppe mammoths. This suggests that a population of *M. meridionalis* may have been ancestral to the steppe mammoth (or at least been modified into an as-yet-unknown species that in turn was ancestral to *M. trogontherii*). While the two species overlapped in time for about a half million years, by about 600,000 years ago the steppe mammoth exclusively occupied the habitats once inhabited by its ancestral species.

The steppe mammoth wandered all over the northern hemisphere, even crossing the Bering Strait into North America. When they reached the New World about 1.5 million years ago they moved south into warmer grassland habitats where a population was adapted into a new species, the Columbian mammoth, which ranged southward into the heart of Central America. Once again, the ancestral species became extinct while the descendant spread out.

The most famous descendant of the steppe mammoth, however, was the smaller woolly mammoth. The transition occurred between about 200,000 and 150,000 years ago, but not in the slow-and-steady pattern that might be expected. By looking at the molars of the mammoths paleontologists have been able to see that the change was relatively abrupt. The steppe mammoth is represented by a particular molar type, but the woolly mammoth molar type evolved quickly and stabilized. The tooth forms seem to be static on either side of the transition, and this points to a particular evolutionary pattern called punctuated equilibrium.

Punctuated equilibrium, known as "punk eek" for short, was first proposed by paleontologists Niles Eldredge and Stephen Jay Gould in 1972. Both scientists were invertebrate specialists, and as such they were able to look at patterns in the fossil record with somewhat more detail than vertebrate paleontologists. Many prehistoric invertebrates were prolific, covered in hard external parts that allowed them to be fossilized relatively easily, and represented by so many individuals that many minute comparisons could be made.

Eldredge's work on trilobites, a diverse group of marine arthropods that looked like a cross between a horseshoe crab and a pillbug, tipped off the duo that species did not always evolve at a uniform rate over time. Eldredge saw a general condition of stasis in which there was almost no change at all, followed by the abrupt appearance of new species alongside their ancestral forms. This suggested that populations of organisms were becoming isolated and undergoing rapid change into new species that would overlap in time with the species from which they had evolved.

This idea was controversial from the start, especially among evolutionary theorists who specialized in genetics, like Richard Dawkins and John Maynard Smith, but as paleontologists reviewed their data they found similar patterns. For years scientists had ignored stasis because it simply was not interesting, but Gould and Eldredge's paper insisted that stasis was relevant to evolution. The pattern of mammoth evolution supports this. The evolution of the northern hemisphere mammoths over the past four million years shows a pattern of mammoths spreading out, populations evolving into a new species, and that new species spreading out to eventually occupy the land of their ancestors after a period of overlap.

In the case of the woolly mammoth, the species evolved somewhere in eastern Eurasia and then spread both westward toward Europe and eastward into North America, replacing the larger steppe mammoth. At the height of their distribution during the Pleistocene, woolly mammoth populations formed a continuous belt from southern Spain eastward into the heart of the United States, though this range fluctuated as habitats changed. The woolly mammoth was a grazer, dependent on grassland habitats to survive. Where the plains spread, so did the mammoths.

The continuously shifting range of the Pleistocene mammoths was also dictated by the waxing and waning of the world's glaciers. The world has seen many periods marked by the advance of ice sheets and glaciers, but the most recent cycle began about 2.5 million years ago, when ice sheets began to creep over the northern hemisphere. Since that time, they have advanced and retreated according to a 40,000- to 100,000-year cycle, with this pattern constantly opening and closing pathways for mammoths. When the glaciers built up the sea level dropped, opening coastal routes to otherwise inaccessible feeding grounds; when the ice melted sea levels rose and flooded those routes. Not only did this cycle influence where mammoths roamed, but it also led to the evolution of pygmy mammoths.

Around 47,000 years ago, glacial advance dropped sea levels enough to reduce the distance between the Channel Islands off California and the mainland to a distance of about four miles, a stretch that the mammoths apparently swam across just as Asian elephants can swim for miles out in the sea today. The mammoths that took the brisk dip eventually made it to the island of Santa Rosae, but as the glaciers melted sea level rose and increased the distance between the island and the California coast, making it too far for the mammoths to swim back. Rather than die out, however, the mammoths became adapted to their new island home by becoming dwarfed. How this occurred is still being debated by scientists, but the island mammoths eventually stabilized at no more than seven feet high at the shoulder, half the height of the ancestral species. They were distinct enough to receive their own species name *Mammuthus exilis*, and they persisted on that island until about 11,000 years ago.

Other populations of mammoths and elephants underwent similar changes. There was a collection of different species of dwarf elephants on islands presently submerged in the Mediterranean, islands containing miniscule representatives of the genus *Stegodon* in Indonesia, and miniaturized woolly mammoths on St. Paul Island off Alaska and just across the Bering Strait on Wrangel Island, just north of Siberia. The causes that led to the creation of the island mammoth off California probably led to the origin of these species as well, and some of them survived for a long time on their island refuges. The St. Paul Island mammoths survived until about 8,000 years ago and the Wrangel Island mammoths until about 4,000 years ago, the latter date being around the time that the Egyptian empire was flourishing and the alphabet was being developed.

As mammoths spread across the northern hemisphere, other elephants spread to new continents as well. Around the time that *Mammuthus* was beginning to colonize Eurasia, the "splendid isolation" of the South American continent was ended when it was joined to North America via the isthmus of Panama. As continental drift steadily pushed the continents closer together, various species island hopped across the narrowing strip of water between them, but the establishment of the land connection about four million years ago opened a thoroughfare for creatures to strike out for new territories. From the south came creatures like giant ground sloths and their tanklike glyptodont cousins, and the North American mastodons *Stegomastodon* and *Cuvieronius* (along with numerous other creatures) pushed south.

With the exception of Australia and Antarctica, during the Pleistocene there were numerous genera and species of elephants on every continent. Some of these, such as the American mastodon, were the last stalwarts of much older lineages, while other elephants, especially the mammoths, rapidly speciated and spread far and wide. But by about 10,000 years ago, most of these creatures were almost, if not entirely, extinguished, including the American mastodon, the woolly mammoth, *Stegomastodon* and *Cuvieronius* in South America, and many of the dwarfed island elephants. Only the ancestors of the modern African and Indian elephants, the "tattered remnants" of elephant diversity, remained.

Nor were the mammoths and mastodons the only species to disappear. Many of the world's largest animals, from saber-toothed cats to giant ground sloths, all went extinct at around the same time. Even the

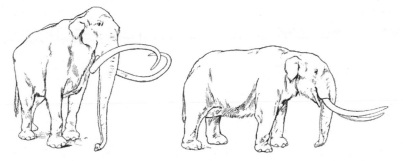

FIGURE 76 – Restorations of a Columbian mammoth (left) and an American mastodon (right), two of the elephants which lived in North America until about 10,000 years ago.

large mammals that survived, such as lions and horses, had their ranges severely restricted. What could have killed so many unique Pleistocene animals and spared the creatures we are familiar with today?

As with any murder mystery, identifying the victims and their time of death is important, and each continent that was home to the lost megafauna of the Pleistocene tells a different story. In North America about thirty-four large mammals, in addition to a handful of smaller ones, disappeared during an extinction spike between about 16,000 and 10,000 years ago. Though the timing has so far been difficult to pin down for South America, it appears that many of their large mammal species were gone by 10,000 years ago, as well.

The pattern in Eurasia was more complex. Between about 49,000 and 23,000 years ago many of the large mammals adapted to warm climates—such as the straight-tusked elephant *Elephas antiquus* and hippos—became extinct. As glaciers advanced so did cold, dry grasslands that were home to grazers like mammoths and horses. Yet some of these large steppe mammals themselves went into sharp decline as of 14,000 years ago, eventually dying out.

Australia, while it was never home to any elephants, also lost species. The island continent was home to many large marsupials unlike anything seen anywhere else in the world, like the wombat relative *Diprotodon*, the size of a small car, and a leopard-sized carnivorous marsupial named *Thylacoleo*, which had shearing teeth shaped like meat cleavers. The extinction event that claimed them kicked off earlier, about 40,000 years ago, with different groups tapering subsequent into nonexistence.

The only exception to the general pattern appears to be Africa. The continent has lost some large mammals within the last 100,000 years and yet it still is home to many of the large mammal groups that where decimated elsewhere.

Numerous non-mutually-exclusive explanations for the great dying during the late Pleistocene have been proposed over the years. Despite recent headlines about how a swarm of comets that struck North America about 12,900 years ago may have been the extinction trigger, most of debate has centered over the influence of two major known killers: climate change and humans.[69]

Climate change has been the traditional suspect. Around 20,000 years ago the world was marked by the Last Glacial Maximum, or the further reach of the ice sheets before they began to recede again. After this time, the warming climate peeled back the ice from the continents, causing sea levels to rise, and forests grew where cold grasslands once dominated. This rapid climate change established the present inter-glacial period, generating the relatively warm and wet world we presently inhabit. As we have recently become aware due to the effects of human-caused climate change, rapid swings in climate can put large animals adapted to cold habitats at risk of extinction. Large mammals in the past would likewise have been affected by rapid change.

In the case of the mammoths, their success was largely tied to the spread of grasslands, and after 20,000 years ago that grassland habitat was shrinking. Their preferred haunts became more restricted to areas like northern Siberia, and fragmentation of their habitat caused the flow of genes between populations to be cut off. The same pressures would affect other grazers that shared the same steppe habitats, and the predators of the large megamammals would find their food sources restricted as the herbivores dwindled.

Global climate change would seem to be a nearly made-to-order cause for extinctions that would affect a large number of animals all over the world, yet it is not without some significant problems as an extinction trigger. One snag is that many of the mammals that became extinct had survived similar fluctuations before. It is difficult to explain why they would have survived previous disruptions but perished during the most recent. Nor was it just the cold- adapted grazers that became extinct. The American mastodon, a browser who lived in more forested environments, its South American cousins, and the Columbian mammoth, which had evolved to live in more temperate climates—all became extinct, too. How climate change would have bent their habitats to the breaking point is unknown, and at present our understanding of the impact of climate change does not entirely fit the global pattern of extinction.

In recognizing the habit of our species to alter the environment in ways detrimental to large mammals, though, some researchers have preferred a scenario in which humans took a direct role in eliminating the world's megafauna. Though *Homo erectus*, Neanderthals, and other prehistoric humans inhabited Eurasia over the last two million years, it was *Homo sapiens* that became a true world traveler, leaving Africa about 100,000 years ago and spreading all over the world.

The expansion of our species from Africa, through Eurasia, and onto other continents appears to fit the pattern of extinction. Whereas extinctions were not as severe in Africa and Eurasia, where different species of humans had lived alongside large mammals for thousands to millions of years, the extinctions in Australia, North America, and South America appear to have occured when humans took up residence on those continents. Waves of migration might have pushed a few humans to these places during earlier times, as suggested by an over 14,000-year-old stone tool found in an Oregon cave discovered in 2009, but it was not until larger populations of humans became established in North and South America by around 11,000 years ago that most of the large mammals disappeared.

The questions are whether this correlation signals causation and by what means humans could have eliminated the megamammals. Hunting has been the most commonly considered factor, especially since the bones of extinct elephants and stone tools have regularly been found together at archaeological sites. Humans may have hunted large mammals into extinction, and those they did not kill directly may have been extirpated by other human influences such as disease, pests, and parasites that traveled with them to new places.

This idea certainly has popular appeal that conflicts with the notion that early humans lived in harmony with nature, but as with the rapid climate change model, the "overkill hypothesis" has some major problems. Among them is the fact that definitive signs that humans hunted large mammals, such as a series of hunting spears lodged in the bones of a mammoth, are exceedingly rare. Relatively more common, but still hard to come by, are bones showing signs that humans butchered the carcasses of some prehistoric elephants for meat. Faced with this association, however, we cannot be entirely certain that the humans hunted and killed that animal or how often they did so.

Looking at the associations of fossil elephant bones and human stone tools in Eurasia, especially, human butchering of mammoth carcasses is rare and usually seen in solitary skeletons. Nor can it be determined for certain whether humans hunted the animals actively or scavenged their carcasses. The somewhat surprising lack of signs of cut marks on mammoth bones, even at sites where multiple mammoth skeletons are preserved, suggests that perhaps mammoth meat was not nearly as important to some humans as that of other, easier-to-hunt animals like deer and bison.

There are also signs that humans butchered mammoth carcasses in North America, but it cannot be taken for granted that these associations indicate hunting on the part of humans, much less a level of hunting so severe as to wipe out the species. It is also strange that, if humans were the main cause for the extinctions of mammoths, the extinction of the mammoths was relatively slow, with several populations that persisted for thousands of years after the 10,000 year mark.[70] In fact, the signs that humans actively hunted and killed many other large mammals that went extinct are either extremely rare or nonexistent. The pattern of human occupation and large mammal extinction seems to fit, but this correlation might not fully explain the ecological catastrophe.

Though arguments over climate change versus overkill can still become fierce very quickly, a more nuanced approach is beginning to emerge. Rapid climate change surely would have affected large mammals, especially those adapted to cold grassland habitats, and as this change was occurring our species was spreading across the planet. The habits of our prehistoric relatives, especially if they regularly hunted large mammals, may have pushed some of the largest mammals over the edge into extinction. The loss of these keystone species might have sent through already destabilized habitats a cascade that precipitated further extinctions.

Some ecologists have proposed that the lost Pleistocene fauna could be restored One such proposal, formally outlined in a 2005 article in the journal *Nature*, is called "Pleistocene Rewilding." Conceived and supported by those who believe that humans are largely to blame for the extinction of large mammals in North America, if not the rest of the world, this project would establish fenced-in parks or reserves in which large mammals would be allowed to roam free. African and Asian elephants, lions, cheetahs, horses, and other animals would be used as proxies for extinct species and mixed with modern bears, pronghorn, and other endemic animals.

This proposal has generated a lot of controversy but relatively little understanding. While its supporters are not seeking to unleash elephants into the suburbs of the American West, it contains some fundamental conceptual flaws. For example, some extinct mammals may have relatively close living relatives; others, such as giant ground sloths, do not. Without the latter a late Pleistocene wilderness would not be complete. Likewise, as a result of climate change (both natural and human caused), some of the habitats—such as the widespread steppe favored by mammoths—the lost animals would have occupied are now lost. Neither does evolution stop to wait for extinct animals to catch up. Organisms and ecosystems have continued to evolve, and there is no certainty that an artificial collection of mammals would interact with modern environments as their ancestors did. The main motivation behind "Pleistocene Rewilding" is penance for what our Pleistocene relatives did; while it might be an interesting ecological experiment, the project cannot turn back evolution to a time before many of the large mammal species were lost.[71]

But what if some of the extinct species could be resurrected with the help of preserved DNA? In an April 1984 issue of MIT's *Technology Review*, a curious article appeared entitled "Retrobreeding the Mammoth," which claimed that, with the help of MIT scientist James Creak, the Siberian veterinarian Sverbighooze Nikhiphorovitch Yasmilov had managed to rejuvenate some eggs recovered from a frozen female mammoth, inseminate them with Asian elephant sperm, and bring two of the embryos to term in female Asian elephants. The resulting offspring, indistinguishable from woolly mammoths, were to be called *Elephas pseudotherias*, and Yasmilov had big plans to breed them to help rescue stranded Siberians workers or help repair the Trans-Siberian oil pipeline.

Newspapers ate it up, but most neglected to check the date of the article or check with the *Technology Review* about the story. It had been an April Fool's joke. Nevertheless, the idea of cloning a mammoth has remained popular. After all, scientists have found a number of well preserved woolly mammoths frozen in the ice of Siberia over the years. If we have the intact hair, skin, and organs of the animals how hard could they be to clone?

The novel *Jurassic Park* and its film adaptation, in which dinosaurs are cloned from blood preserved inside prehistoric mosquitoes trapped in amber, have made the idea of bringing an extinct creature back to life seem like a piece of cake. All you need is some preserved DNA and—bingo!—the past comes alive. Anyone who was paying close attention to the sci-fi story will have noticed that the fictional scientists glossed over some crucial parts of the process, however. It is not so easy to go from a string of ancient DNA to a living organism.

Extracting DNA from mammoth skin, hair, or teeth would just be the initial step in a very complicated process, made all the more difficult

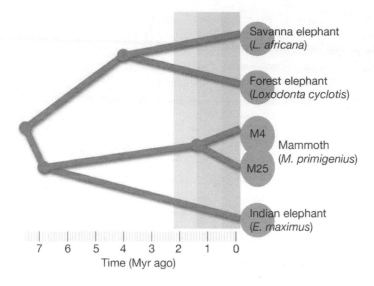

FIGURE 77 – An evolutionary tree of living elephants and woolly mammoths based upon genetic data. Asian elephants are more closely related to woolly mammoths than either is to African elephants.

by the fact that DNA degrades over time. Scientists cannot simply pluck a cell from a frozen mammoth and map its entire genome. The entire complement of DNA would have to be carefully identified and put together by looking at multiple specimens. To date, scientists have identified the complete sets of mitochondrial DNA for woolly mammoths and the American mastodon, or the unique DNA found in that sausage-shaped organelle inside the cell, but we are still a long way from mapping the entire set of DNA that would be contained in the cell nucleus, the necessary genetic material for the cloning of an animal.

For the sake of argument, however, let's assume that we presently have the complete set of nuclear DNA from a woolly mammoth. How do we get from that to a cute, fuzzy newborn mammoth? The fact that genomes might be compared to blueprints or recipes does not explain how they get translated into a living organism.

What we would need to do is create a fertilized egg that can grow inside the womb of a mother animal. This would involve removing the DNA from the unfertilized egg of an Asian elephant, fertilizing that with revitalized mammoth sperm (the DNA inside of which would probably be degraded), and getting that to grow to term inside a female Asian elephant. The embryo would require such an environment in which to develop and grow; it cannot be replicated outside the womb. All of this, of course, is an oversimplification. This process would be unimaginably intricate and require an knowledge of DNA, ancient DNA no less, that we do not as yet possess. And even if it could be accomplished, we must ask whether the end result would be a true mammoth, especially if input from the Asian elephant, the woolly mammoth's closest living relative, was required at every single step.

Perhaps there will come a day when cloning a mammoth will be possible, but for now it seems that it would be easier to breed the hairiest Asian elephants together over many generations until you had individuals with long, heavy coats. It would not be a mammoth, but it would be about as close to seeing one again as we could get. Despite the low likelihood that we will be able to clone a mammoth (even cloning living animals is still fraught with problems) attempts at doing so could still teach us much about the recovery and study of prehistoric DNA. Perhaps that is not as glamorous as producing a species that has been extinct for thousands of years, but it would help us better understand how life has evolved and continues to do so.

On a Last Leg

"It has become evident that, so far as our present knowledge extends, the history of the horse-type is exactly and precisely that which could have been predicted from a knowledge of the principles of evolution."
—Thomas Henry Huxley, *American Addresses*, 1877

On February 9, 1857, Emily Gosse passed away. Suffering from a breast cancer discovered only a few months before, her health had ebbed bit by bit until she succumbed to the disease. It was a devastating loss for her son, Edmund, and her husband, the celebrated naturalist Philip Henry Gosse.

Several months prior to Emily's death, Gosse seemed to be at the high point of his career. His highly regarded 1854 book on marine biology, *The Aquarium*, had inspired a frenzy of aquatic-themed English decoration, and led to his election as a Fellow of the Royal Society in 1856. Emily's death made this academic victory seem sour. Gosse's despair deepened when his hastily written biography of his wife, *Memorial of the Last Days on Earth of Emily Gosse*, was not well received by his friends and family. The now widowed and middle-aged Gosse felt isolated, and the departure of his wife required that he ensure the Christian salvation of his son at a time when the ruminations of evolutionists were becoming increasingly prominent. The incessant hammering of geologists and the macabre activities of comparative anatomists were threatening the ideals Gosse believed in most fervently.

Figure 78 – Philip Henry Gosse and his son, Edmund, photographed in 1857.

His next two publications were direct attacks on evolution, a threat he perceived as remaining partially concealed from the public. The first book, which appeared in 1857, was a collection of essays he had written for the magazine *Excelsior*, entitled *Life in its Lower, Intermediate, and Higher Forms: or, Manifestations of the Divine Wisdom in the Natural History of Animals*. As the title suggested, Gosse had a hierarchical vision of the natural world, in which God fitted each organism to a purpose. His son would later admit in a biography that Gosse's approach in this volume was more heavy-handed than usual:

> These essays were slight, and the religious element was quite unduly prominent, as if vague forebodings of the coming theory of evolution had determined the writer to insist with peculiar intensity on the need of rejecting all views inconsistent with the notion of a creative design. This book entirely failed to please the public, who had now for so many years been such faithful clients to him; with the scientific class it passed almost unnoticed.

In Gosse's next book, *Omphalos*, published the same year, he proposed to once and for all reconcile geology with Genesis. While many before him had tried, Gosse believed that he had the ultimate answer to the conundrum. He declared that the entire expanse of geological time, from the days mammoths trumpeted on the Siberian steppe to the era dominated by the fearsome dinosaurs and beyond, was just an illusion.

If there was any conflict between science and the Bible, Gosse argued, science was in the wrong. The world had been created with the *appearance* of age. According to Gosse, the strata of the earth were analogous to the rings inside the mature trees that God had created. God knew that each living thing had a life cycle and he merely chose the adult stage to start with, thus creating the world and the creatures within it according to the foreordained trajectories of development.

Theologians and naturalists alike were dumbfounded by Gosse's thesis. To naturalists, *Omphalos* appeared as an anachronistic attempt to shoehorn all of nature into the confines of Gosse's interpretation of Genesis; to the faithful it cast God as a trickster. Even those who were sympathetic to Gosse's aims, if not his arguments, could not accept his views. In a personal letter to Gosse, his clergyman friend Charles Kingsley charged that, in proposing the world to have been created with the impression of great antiquity, "you make God tell a lie."

Gosse was crestfallen, his views dismissed and forgotten by the time Charles Darwin's *On the Origin of Species* was published in November of 1859. Nature would be understood on its own terms. Lamarck's hypothesis had been sneered at, Robert Chambers's anonymous tract had been dismissed, and the work of Wallace had been pushed aside. Yet while social conservatives such as Richard Owen and Adam Sedgwick bridled at Darwin's evolutionary vision, *On the Origin of Species* brought the debate over the evolution of life to a higher level.

Despite the massive amount of data Darwin mustered in defense of his ideas, however, the lack of transitional fossils remained the most often-repeated objection to his hypothesis (perhaps because, in his honest treatment of the subject, Darwin identified it himself). Among those who doubted Darwin's mechanism was the French paleontologist Albert Gaudry.

Between 1855 and 1860, Gaudry participated in the excavation of a Miocene fossil site in Pikermi, Greece, that contained the bones of mammals of an age intermediate between those already known from the older Eocene and the much younger Pleistocene. If Darwin was correct, then the Pikermi strata would contain mammals intermediate in form between those of the earlier and later periods, and that is precisely what Gaudry found. There were fossil monkeys, giraffes, rhinos, elephants, saber-toothed cats, and more, many of which appeared to fill in a gap in prehistory. As Gaudry would later write:

> Thanks to the palaeontological researches which are everywhere conducted, beings of which we did not understand the place in the economy of the organic world are revealed to us as links in chains which themselves are connected; one finds transitions from order to order, from family to family, from genus to genus, from species to species.

These transitions fit within the evolutionary framework Darwin had predicted. When *On the Origin of Species* was published, it contained only one illustration, a "tree of life" that revealed evolution as a gradual, branching process. Darwin's tree was only hypothetical, but Gaudry was one of the first to take this general concept and plug in actual organisms in order to trace their "filiation" (or relatedness to each other). One such tree, published in 1866, represented the evolution of horses. At the time, horses were recognized perissodactyls, or ungulates with an odd number of toes, but how they were related to the rest of the group was difficult to determine. Even though it had first been described years earlier, for Gaudry the equid *Hipparion*, discovered at Pikermi, provided both an anatomical and a temporal intermediate between the more recent horses and older, more generalized perissodactyls. *Hipparion* differed from modern horses in that it had small toes on either side of the hoof that did not touch the ground, thus connecting the one-toed members of *Equus* with other perissodactyls, such as tapirs and rhinos, that bore more toes on their feet. This, of course, still left the problem of what animal *Hipparion* itself arose from, but it was a start in filling out the history of horses.

The fossils of *Hipparion* found at Pikermi were important for another reason related to natural selection, as well. There were enough of them to study the variation of the genus. Just like there are differences between individuals of any one population or species living today, so there is variation between the remains of fossil animals, and variation is the raw material that natural selection works on to produce evolutionary change. In fact, the remains of *Hipparion* exhibited so much variation that Gaudry noted that they may have been considered distinct species had they not all been found together, which in turn suggests that gaps between gaps will closed up as more fossil material is recovered.

Yet Gaudry did not believe that natural selection could account for the evolutionary patterns he saw. Much as Gaudry appreciated Darwin's making biological transformation a legitimate question, he dismissed natural selection because he abhorred the idea that evolution could involve an element of chance:

> If we recognise that organised beings have little by little been transformed, we shall regard them as plastic substances which an artist has been pleased to knead during the immense course of ages, lengthening here, broadening or diminishing there, as the sculptor, with a piece of clay, produces a thousand forms, following the impulse of his genius. But we shall not doubt that the artist was the Creator himself, for each transformation has borne a reflection of his infinite beauty.

Regardless of his disagreements with Darwin, however, Gaudry's work on the placement of *Hipparion* as an intermediate and possible ancestor to the modern horse was only a relatively early example of the importance of horses in illustrating evolutionary change.

It would not be long before other paleontologists picked up where Gaudry had left off. During the 1870s a Russian émigré and former student of Ernst Haeckel named Vladimir Kovalevskii rifled through cabinets of bones in Parisian museums, including Cuvier's collections from around the city itself. Within these collections lay the the bones of *Anchitherium*, which had existed in just the right place and time to bridge a major gap in horse evolution, and the correlation of features from *Palaeotherium* on the end through *Anchitherium* and *Hipparion* to *Equus* seemed to suggest adaptation to changing environments. Living horses were known to be grazers, with high-crowned teeth adapted to eating grass. *Palaeotherium* and *Anchitherium* had lower-crowned teeth that were much more similar to those of browsers, like tapirs, and so Kovalevskii, who did accept Darwin's mechanism for evolution, not only had a transitional series but also a trigger for the mechanism of natural selection to act. (Although his scientific work on the evolution of horses was significant, Kovalevskii was never welcomed into professional paleontological circles and, beset by depression and financial woes, he committed suicide by chloroform in 1883.)

Kovalevskii's general series was very similar to the linear transition of horses that T. H. Huxley presented in his 1870 address to the Geological Society of London. Huxley's address was primarily concerned with presenting evidence from paleontology for evolution, including representative types of the transition from reptiles to birds and from terrestrial carnivores to whales. In both cases, however, fossil remains were only representative of the probable way in which evolution proceeded. Horses, however, presented a direct line of descent, and as such were a perfect example of how evolution was affected through successive modification.

> When we consider these facts, and the further circumstance that the Hipparions, the remains of which have been collected in immense numbers, were subject, as M. Gaudry and others have pointed out, to a great range of variation, it appears to me impossible to resist the conclusion that the types of the Anchitherium, of the Hipparion, and of the ancient Horses constitute the lineage of the modern Horses, the Hipparion being the intermediate stage between the other two.

This three-step path from *Anchitherium* to *Equus* would only be the start. The theory of Darwin and Wallace predicted that there would be an even older equid that would connect *Anchitherium* to an even more ancient ancestral type, and Huxley felt that a small creature called *Plagiolophus* (originally called *Palaeotherium minor* by Cuvier) fit the role nicely. Although cautious about placing the little perissodactyl in the direct line of descent, Huxley still felt that it was at least representative of the form of the creatures from which later horses evolved.

The evolution of the horse thus seemed almost completely understood. Even after Huxley began to shift his research interests away from paleontology, horses remained a particularly good example of evolution. The fossil horses were the stars of his 1876 lecture tour of the

United States, where Huxley summarized the views he had earlier presented to the Geological Society:

> Seven years ago, when I happened to be looking critically into the bearing of palaeontological facts upon the doctrine of evolution, it appeared to me that the *Anchitherium*, the *Hipparion*, and the modern horses, constitute a series in which the modifications of structure coincide with the order of chronological occurrence, in the manner in which they must coincide, if the modern horses really are the result of the gradual metamorphosis, in the course of the Tertiary epoch, of a less specialized ancestral form. . . . That the *Anchitherium* type had become metamorphosed into the *Hipparion* type, and the latter into the *Equine* type, in the course of that period of time which is represented by the latter half of the Tertiary deposits, seemed to me to be the only explanation of the facts for which there was even a shadow of probability.

The procession of forms made sense in terms of anatomy and fossil sequence, but a new set of discoveries would overturn what had been Huxley's prime example of evolution. These would be made not in the Old World, but in the New.

By the middle of the nineteenth century some European paleontologists were becoming worried about the state of their science. It seemed as if most of the available fossil strata had all been thoroughly investigated, and even these were separated by gaps in the geological record that could not be found in Europe. The German paleontologist Ferdinand Roemer even went as far as to tell his students to avoid populated areas and seek fossils in wilder, unsettled regions. This point was apparently not lost on one of his students, O. C. Marsh.

Marsh's professional career got off to an unusual start. Through the connection to his wealthy uncle, the entrepreneur George Peabody, Marsh was able to wrangle a professorship in paleontology at Yale. It would be the first such official post in the nation. The college did not actually have the money to pay Marsh, but Peabody endowed the school with funds to build a museum for his nephew, and provided him with an

FIGURE 79 – Paleontologist O. C. Marsh.

allowance that kept him living comfortably. Since he was not being paid by Yale, Marsh had no teaching duties or other responsibilities placed upon him. He was free to pursue his professional career as he saw fit.

During the time Marsh was studying anatomy and paleontology in Germany prior to his Yale appointment, the general consensus was that there was no strong evidence that ancient horses had ever lived in North America. Recalling such statements Marsh later wrote:

> I heard a world-renowned professor of zoology gravely inform his pupils that the horse was a gift from the Old World to the New, and was entirely unknown to America until introduced by the Spaniards. After the lecture, I asked whether no earlier remains of horses had been found on this continent and was told in reply that the reports to that effect were too unsatisfactory to be presented as facts of science. These remarks led me, on my return, to examine the subject myself.

The opportunity for Marsh to prove this professor wrong came soon enough. In 1868, while traveling to the end of America's unfinished transcontinental railroad, he heard that some strange bones had been recovered from a well in nearby Antelope Station, Nebraska. Smelling an opportunity, Marsh tipped the conductor to hold the train while he went to see if the rumors were true, and soon found what appeared to be the bones of diminutive horses. Marsh named his "toy" horse *Equus parvulus*, later renamed *Protohippus*.

This was just the beginning of Marsh's fossil horse discoveries. Accompanied by teams of students (and sometimes cavalry escorts) Marsh collected the bones of numerous fossil horses from almost the entire geological span of the previous fifty-five million years. These were all brought back to the Peabody Museum at Yale for study. Marsh had gathered a much more complete and detailed sequence than the lineage Huxley proposed in 1870.

Marsh was certain he had found the direct line of descent from small, many-toed ancestral horses to large, single-toed descendants. The horse pedigree ran from the four-toed *Orohippus* of the Eocene through the three-toed *Mesohippus* to the Miocene forms *Miohippus* and *Protohippus* and to the one-toed Pliocene *Pliohippus* before culminating in the familiar *Equus* of the Pleistocene on. It was a simple, beautiful sequence that left no room to doubt that horses had evolved in North America. The sequence preferred by Gaudry, Kovalevskii, and Huxley only documented migrants that had left the continent of their birth.

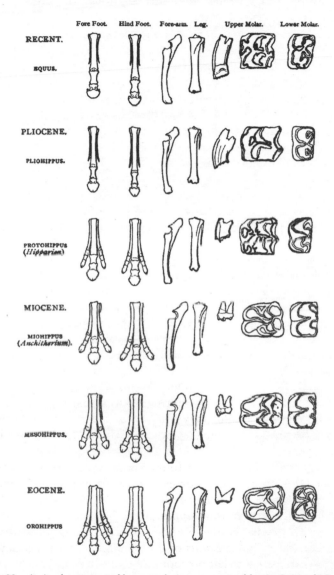

FIGURE 80 – A visual summary of horse evolution as proposed by O. C. Marsh. In Marsh's diagram, there are clear, unidirectional trends in toe reduction, tooth height, and, though it cannot be seen here, size.

Huxley was faced with this wall of evidence when he met with Marsh during his lecture tour and could do little except discard his shorter European horse genealogy for Marsh's more comprehensive one. For every question Huxley had, Marsh seemed to have a bone in response. As he wrote to his wife of the meeting, Huxley "turned upon [Marsh] and said: 'I believe you are a magician; whatever I want, you just conjure it up.'" Huxley was most taken with the four-toed *Orohippus*, and predicted that a five-toed horse would eventually be discovered in even older Eocene deposits. (In jest, Huxley drew a cartoon of the elusive "*Eohomo*" riding the "Dawn Horse".)

Huxley quickly changed his lecture to include the new information. During his American lecture tour he argued Marsh's case for a North American origin of horses, and Huxley concluded:

> Thus, thanks to these important researchers, it has become evident that, so far as our present knowledge extends, the history of the horse-type is exactly and precisely that which could have been predicted from a knowledge of the principles of evolution. And the knowledge we now possess justifies us completely in the anticipation that when the still lower Eocene deposits and those which belong to the Cretaceous epoch, have yielded up their remains of ancestral equine animals, we shall find, first, a form with four complete toes and a rudiment of the innermost of first digit in front, with probably, a rudiment of the fifth digit in the hind foot; while, in still older forms, the series of the digits will be more and more complete, until we come to the five-toed animals, in which, if the doctrine of evolution is well founded, the whole series must have taken its origin.

Although neither Huxley nor Marsh was aware of it at the time, the remains of tiny *Eohippus* were with them as they looked over the remains of its younger relatives. As Marsh wrote to Huxley two months later, "I had him 'corralled' in the basement of our Museum when you were there but he was so covered with Eocene mud that I did not know him from *Orohippus*. I promise you his grandfather in time for your next horse lecture if you will give me proper notice." With four front toes and the vestiges of a fifth, the tiny horse relative corresponded almost exactly to what Huxley had proposed would be found, and this creature corresponded closely to another fossil mammal that had been described years before under the name *Hyracotherium*.

In 1832, William Colchester found two fragments of a fossil jaw in Woodbridge, England. They were brought to the attention of Richard Owen, who supposed that they had belonged to some kind of monkey,

FIGURE 81 – T. H. Huxley's cartoon of an "Eohomo" riding his vision of what an "Eohippus" might look like.

but as more fragments from similar deposits were found, including a partial skull, Owen realized that the animal was not a monkey at all. He described it under the name *Hyracotherium* in 1839, but just what type of mammal it was remained unclear. In life it would have had "a resemblance to [the physiognomy] of the Hare, or other timid Rodentia," though in affinity Owen thought *Hyracotherium* was probably a member of the "natural family of the Hog tribe." He even described a second species of the "rabbit-like" *Hyracotherium* in 1843 on the basis of several teeth, but the discovery of *Eohippus* allowed the affinities of *Hyracotherium* to be understood. The small creature was no hog relative at all but was more closely related to the earliest known horses, although naturalists disagreed as to whether *Hyracotherium* or *Eohippus* should be the proper name for this early form.

Unfortunately Marsh was not able to make good on his promise of an even earlier horse ancestor with five complete toes, but by illustrating the graded evolution of the horse he had unquestionably verified what Darwin's theory predicted. Darwin himself appreciated the evidence that Marsh had provided his theory, noting that the paleontologist's work "afforded the best support to the theory of evolution which has appeared in the last 20 years."

The evolutionary series seemed to be complete. There were so many horse fossils that it was easy to create displays showing how horses had gone from small, generalized animals to the powerful beasts of burden we know today. Marsh's "direct line of descent," however, would not hold up against the mounting wave of horse fossils that continued to come in from the western United States. There were simply too many extinct species to array in a single straight line.

In a 1907 review of fossil horses, the paleontologist G. W. Gidley surveyed the fossil horses from the previous thirty-four million years and found that many of these multi-toed fossil horses were not directly ancestral to living forms. In fact, at any point in time there seemed to be as many as four genera of horses living alongside one another. He envisioned the evolution of horses in a branching pattern, with each family being self-contained. Yet, even if the general patterns were apparent, the exact connections between types were still difficult to pin down.

Walter Granger complemented Gidley's work in 1908 with a revision of the geologically older Eocene horses of North America that formed the base of the later radiation Gidley had cataloged. These earlier fossils were especially problematic, as they were close to the ancestry of not just horses, but also of the massive, horned brontotheres and other kinds of perissodactyls as well. Virtually the entire swath of perissodactyl diversity had been derived from very similar early types that resembled *Hyracotherium*, thus placing the branching pattern of horses into an even more tangled evolutionary pattern.

These more detailed problems persisted even after the efforts of Gidley and Granger. In 1924 W. D. Matthew, another scientist in H. F. Osborn's stable at the American Museum of Natural History, published a paper entitled "A New Link in the Ancestry of the Horse." In his introduction Matthew wrote:

> The series of American Tertiary ancestors of the horse is one of the classic examples of evolution provided by the fossil record and the most complete and convincing among the mammals. Nevertheless, it is well recognized by those who have made a special study of it, that, while the broader lines of descent are beyond reasonable question, there are definite gaps between some of the successive stages and many minor problems as to the details of phylogeny.

One such gap was between the late Pliocene horse *Pliohippus* and the Pleistocene genus to which modern horses belong, *Equus*. While *Pliohippus* was large and the toes on the lateral sides of its feet had been reduced practically to nubs, it still had some peculiar features, such as a pit on the either side of its face, which distinguished it from modern horses. There had to be an intermediate form between it and *Equus* if it was the true ancestor of modern horses. Matthew believed that he had found such an animal. He called it *Plesippus*, and while *Pliohippus* would later turn out to be further removed from the ancestry of modern horses than had been previously thought, *Plesippus* did possess transitional features that linked it between archaic horses and the earliest *Equus*.

Even as paleontologists recognized the branching pattern of horse evolution, however, the linear iconography favored by Marsh held fast. The tension between these ideas was embodied by a review of horse evolution Matthew published in 1926. In general, it seemed apparent that horses got bigger, lost their "extra" toes, developed elongated faces, and evolved high-crowned teeth fit for grazing all at the same time. The traits were not obtained one at a time, but were all gradually modified in one direction. This could clearly be seen in the updated evolutionary trajectory Matthew published (*Eohippus–Orohippus–Epihippus–Mesohippus–Miohippus–Parahippus–Merychippus–Pliohippus–Plesippus–Equus*) which was taken as representing the successive "stages" of horse evolution. From oldest to youngest, each genus represented a "step forward" for horses. In this view, even if we had only the skeletons of the beginning (*Eohippus*) and end (*Equus*) of the series it would still be possible to predict all the intermediates along the graded chain.

This progression was in stark contrast to the full diversity of fossil horses Matthew recognized. In a more technical diagram, representing the span of time each genus was found and their connections to each other, it was clear that as many as seven different genera overlapped in time. Ancestors even persisted alongside their descendants in some

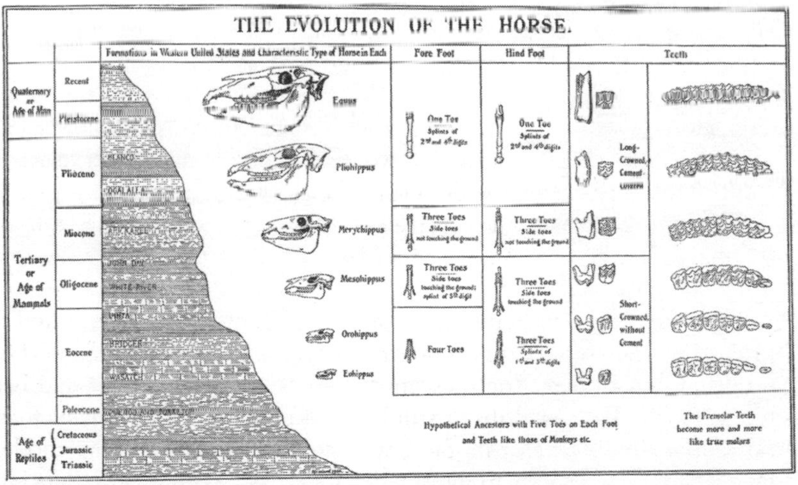

FIGURE 82 – A simplified version of horse evolution as forwarded by W. D. Matthew. Even though Matthew recognized that horse evolution was more "bushy" with different groups overlapping in time and many not being ancestral to living forms, the most popular visualization of horse evolution was this modified form of Marsh's diagram.

cases, but Matthew did not pay much attention to this. He lumped most of what were considered "side branches" under the heading "Divergent Lines of Equidae" in this paper, implying that they diverged from the "main line" of evolution from *Eohippus* to *Equus*. Hence *Hipparion* was presented as a dead-end offshoot, as was a South American form with an enigmatically large nasal cavity called *Hippidium*, and many others. For Matthew, the "main branch" horses had given rise to radiations of "divergent lines" at various times during history, and any genus that expired without pushing the next "stage" of horse evolution forward could be pushed to the side.

Even so, Matthew was more amenable to natural selection than many of his colleagues. He was more interested in how changing climates and environments affected organisms, and to him it was clear that the evolutionary success of the horses was dependent upon the spread of grasslands. The earliest horses, such as *Eohippus*, browsed on leaves in warm forests during the beginning of the Eocene, but as the global climate became cooler and more arid most horses became adapted to life on the grassland. The reduction of toes, the evolution of high-crowned teeth, and other fluctuating characteristics all seemed to be tied to a shift to life on the plains.

Matthew's emphasis on environment would later be complemented by his student, George Gaylord Simpson, in his 1951 popular summary *Horses*, in which evolution by natural selection was given pride of place despite occasional references to "main lines" of descent. In the classic view, the shift from browsing to grazing had driven the larger trends seen in horse evolution. The larger size, high-crowned teeth, and other adaptations had all been selected for as forests gave way to grasslands, especially in the latter part of the Cenozoic (occurring alongside the evolution of grazing elephants).

The basic story went something like this: by the time *Eohippus* had evolved, western North America was covered in warm forests that provided plenty of soft food for the little horses to eat. The successors of *Eohippus*, such as *Orohippus* and *Mesohippus*, continued this lifestyle, but by about nineteen million years ago the climate was becoming cooler and drier. Grasslands overtook forests, so horses were adapted to life on the plains. This resulted in a proliferation of horses adapted to eat grass rather than leaves, such as *Merychippus* and *Hipparion* (among many others), which exhibited many of the classic horse characteristics. The evolution of high-crowned teeth adapted to grinding grass, long legs tipped in a reduced number of toes for fast running on hard ground, long faces, and other traits all arose from this environmental shift.

Yet, as paleontologists have learned more about horse evolution over the last half century this simple story has become more complicated. The traditional view—that the spread of grasslands sent horses moving through a particular evolutionary pathway in which all their defining adaptations were linked together—does not hold up. Part of this can be understood by looking at some strange mammals only distantly related to horses.

When the non-avian dinosaurs went extinct sixty-five million years ago, South America was an isolated island continent, and a unique array of mammals evolved there. Among the varied assemblage was a group of hooved mammals called litopterns, of which the false llama with a trunk—*Macrauchenia*—discovered by Charles Darwin is the most well known. But *Macrauchenia* was only one of the last of the litopterns, a survivor of the Great American Interchange in which so many native animals were wiped out. Much earlier in the history of the group there were two litopterns that convergently evolved to stand on one toe.

One such litoptern, *Diadiaphorus*, exhibited a mix of characteristics seen in early and late horses. Its skull was like that of *Eohippus*, with a short face and a full set of low teeth indicative of an animal that ate soft plants, but its feet were superficially similar to those of later, larger horses such as *Merychippus*. Like them *Diadiaphorus* stood only on one toe and had two side toes that did not touch the ground. This toe reduction was taken even further by its close relative *Thoatherium*. It, too, had a skull that was more similar to that of *Eohippus* than *Equus*, but *Thoatherium* had only one central toe. The only signs that it had ever had side toes were two tiny nubs along its lower leg, and this astonished paleontologists as even living horses have more prominent vestiges of their side toes (the splint bones). Since both *Diadiaphorus* and *Thoatherium* lived during the Miocene, it was clear that they had

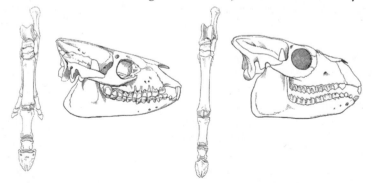

FIGURE 83 – The left hind limbs and skulls of *Diadiaphorus* (left) and *Thoatherium* (right).

evolved more "horselike" legs even before horses themselves did.

It might be tempting to look at such a case of convergent evolution and conclude that the evolution of something horselike is somehow inevitable, but the anatomy of *Diadiaphorus* and *Thoatherium* actually suggest a different conclusion. The two litopterns were clearly not going through some kind of directed sequence from less to more "horsiness" over time. They possessed a mosaic of traits that confirm that the reduction of toes, acquisition of large body size, evolution of high grinding teeth, and an elongation of the face did not have to evolve as a single suite of characteristics. To better understand how such traits evolved in horses, it is profitable to go back to the beginning.

The first step involves identifying the start of horse evolution. This is not quite as easy as it sounds. For over a century, paleontologists have been trying to figure out the relationships between the *Eohippus* fossils of North America and *Hyracotherium* of Europe. The reason for this debate is that the genera are very similar to each other and very close to the base of not only the early evolution of horses, but the radiation of other types of perissodactyls, as well. Early rhinos, tapirs, brontotheres, palaeotheres, and chalicotheres probably all evolved from something akin to what has been classically called *Hyracotherium*.

For many years *Hyracotherium* was used as a taxonomic wastebasket to which hard-to-identify fossils were often assigned. The disarray began to be rectified in 2002 when paleontologist David Froehlich reexamined many of the *Hyracotherium* and *Eohippus* fossils to see whether or not they really did fall under just one or two genera. What he found was that there were more different kinds of early horse relatives than had previously been supposed. While *Hyracotherium* from Europe grouped most closely with creatures like *Palaeotherium,* there were several different early forms at the base of the horse evolutionary tree. The early genera *Sifrihippus, Minippus, Arenahippus,* and *Xenicohippus* were all close relatives of *Eohippus* from North America, placed near the base of early horse evolution.

Eohippus and its close relatives flourished during the time of a global hothouse. Warm climates reached so far north that there were crocodiles living in what is now the Canadian Arctic, and much of North America was covered in subtropical forests. These lush forests provided plenty of food for the small, multi-toed horses of the day, but such succulent plants were not only available in the forest. In 1981, paleontologist Philip Gingerich described a population of two dozen *Hyracotherium* from a site in southern Colorado that appeared to have been a more open, woodland environment. In such a habitat the early

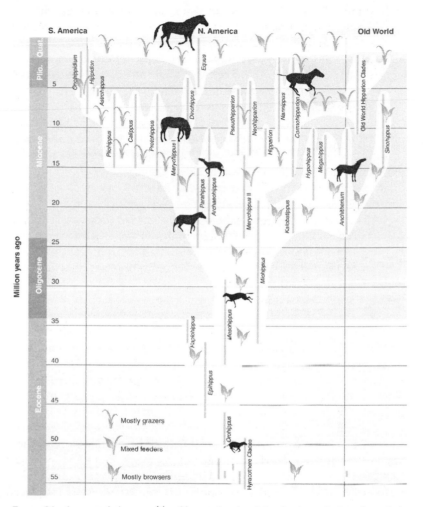

FIGURE 84 – A recent phylogeny of fossil horses showing their adaptive radiations through time and their expansion to different continents.

horses could have picked leaves off shrubs, chewed on ferns, and eaten other soft plant foods that grew low to the ground. That they probably did so has been backed up by studies of microscopic wear patterns on their teeth.

The horses which followed the early *Eohippus* type were not very much different. Horses such as *Orohippus* and *Epihippus* were still relatively small forms that lived in the Eocene hothouse climate in which their ancestors evolved. As the Eocene drew to a close, though, the climate dramatically changed. By about thirty-four million years ago, the average global temperature had dropped approximately twelve degrees Celsius from the Eocene maximun, causing an uptick in extinction that claimed the last of the archaeocetes such as *Basilosaurus*. Horses were not so adversely affected, and as temperatures fell in the late Eocene horses began to change. From forms akin to *Epihippus* evolved *Mesohippus*, a "middle horse" that was subtly different from its ancestors. It was a little larger, its face was a little longer, its teeth had more folds useful for chewing tougher foods, and it supported its weight on three-toed feet. In fact, *Mesohippus* was nearly caught in the act of a transition in that early members of the genus had a still-functional vestige of a fourth toe on their front legs while later representatives did not.

At first this might seem to fit in with the idea that there was a "main line" of horse evolution in which genera gradually transformed from one to another, but a study by paleontologists Neil Shubin and Donald Prothero published in 1989 suggested that this interpretation was wrong. Traditionally *Mesohippus* was treated as a kind of missing link between earlier horses and later horses, a single stock from which the horse *Miohippus* evolved before giving rise to a diversity of later types. Since both genera were well represented from many fossils dug out of the thirty- to thirty-three-million-year-old rock of South Dakota, however, their fossils allowed paleontologists to reappraise this classic view.

Shubin and Prothero's findings grated against the classic story. Rather than just a single lineage of *Mesohippus* gently grading into *Miohippus*, both genera were represented by several species that overlapped in time. There was a radiation of *Mesohippus* species, a population of which was modified into what we call *Miohippus*, and species of this newly derived genus flourished alongside species of *Mesohippus*, which itself did not become extinct until four million years later. Just like the mammoths, populations of species split off, evolved into new species, and lived alongside their ancestral species without much further change (at least until another population speciated, in any case).

Overall, though, *Mesohippus* and *Miohippus* were not all that different from earlier horses. *Miohippus* embodied a shift, with its body weight on a single, central toe (even as the small side toes remained), but it was still a relatively small animal that primarily fed on shoots and leaves. As the Oligocene gave way to the Miocene twenty-three million years ago, however, major environmental changes were occurring. There was an uptick in global temperature, but the subtropical forests gave way to savanna rather than bouncing back. Grasses had been around since the Cretaceous, but it was only now that they began to occupy greater expanses of land, and many mammal groups, elephants and horses among them, responded by becoming adapted to eating grass.

It was a time when horses exhibited the highest amount of diversity. New horse genera radiated across North America, with a maximum of thirteen genera living at the same time about fourteen million years ago. Some of these lineages, such as those containing *Anchitherium* and *Hipparion*, even crossed over into Asia and spread through the Old World.

Chewing grass can be hell on teeth, especially if the grass has little spicules of minerals in it that abrade enamel as the plant food is chewed, like silica. This means that an animal that shifted to grazing would wear down its teeth much faster than a browser would. In contrast to the conveyor-belt dental system of elephants, horses that moved into the grasslands were adapted to have a complete set of longer, high-crowned teeth that would take a lifetime to wear down.

If we were to take an x-ray of the head of a horse, we would see that the crowns of its teeth extend well into the upper and lower jaws. At only five years old the horse would still have much of its complete adult dentition intact. If we were to take another x-ray a few years later, though, we would find that the teeth did not reach as far as they once did. The horse's near-constant chewing wears down the surfaces inside the mouth, so the already-formed teeth are continually pushed up through the gums to make sure there is a fresh grinding surface. Now let us assume that our horse was fortunate enough to live a long life, attaining the age of thirty-five. If we took one last x-ray we would see that the teeth would be almost entirely worn down, with just the roots extending a little way into the skull.

The shift toward these high-crowned teeth occurred about four million years after the spread of grasslands, and as selection favored higher-crowned teeth some changes in the skull were required. Horse faces, or the part of the skull in front of the eye, had generally become longer since the time of the earliest horses, but the lengthening of the teeth required that the faces of grazing horses become even longer. In their ancestral state the last teeth in the horse's upper jaw were right below the eye socket, and as selection favored higher tooth crowns the teeth ran the risk of overcrowding the eye. The selection for higher teeth thus favored the evolution of a longer face by shifting the tooth row forward, a change that otherwise might not have occurred.

The group of horses with these high-crowned teeth belonged to the larger group Equinae, which not only included *Hipparion* and its relatives but the closest relatives of living horses as well. Even though these horses placed their weight primarily upon their central toe many still retained side toes for long periods of time. The side toes did not gradually shrink at a constant rate but remained so prevalent it is almost surprising that there are no three-toed horses today.

Three-toed horses were a casualty of the decline in horse diversity starting about ten million years ago. Among the first to go were the forest-dwelling anchitheres, followed by some of the earliest high-toothed equines such as *Protohippus* and *Pliohippus*. The smaller-sized horses, especially, disappeared, and those already adapted to grasslands either became adapted to the colder, drier steppe habitats that were spreading or went extinct. By about five million years ago there were only about six genera left scattered across the globe, including early members of the genus *Equus*. These were horses of modern aspect, but by one million years ago they were the only horses left. All the browsing horses and three-toed horses were gone, and during the Pleistocene mass extinction that wiped out the mammoths horses were also extirpated from the continent on which they had originated. It would only be a few thousand years later, when Europeans would bring domesticated horses with them to the New World, that horses would return.

A look at the evolution of body size among horses helps to underscore the pattern of their history. Even after ideas about "internal driving trends" were booted out of evolution, it was still believed that horses were a good example of a trend called "Cope's Rule." This is not so much a mechanism as a pattern in which the body size of a group of organisms increases over time.

Certain examples of fossil horses from successive strata would appear to illustrate this trend. The evolutionary lineages created by Marsh, Matthew, and others certainly show it, but the answer lies in an analysis of the entire swath of horse evolution. This task was undertaken by fossil horse expert Bruce MacFadden during the late 1980s, and he found a different pattern. For the first twenty-five million years of horse evolution, prehistoric horses stayed small, estimated at about 50 kilograms or less. It was only after the time of *Mesohippus* that significantly larger horses began to evolve, but there were still relatively small horses through the beginning of the Miocene. About ten million years ago, however, almost all of these small forms were wiped out, leaving horses that had, for the most part, independently evolved to a larger size estimated at 200 kilograms or more. At this time the small, forest-dwelling horses that had existed for about forty-five million years were decimated. The only exceptions were lineages such as the Pliocene horse *Nannippus* that actually became dwarfed in its evolution from a larger ancestor.

The reduction of toes, another favorite textbook trend, is similarly complex. Early horses, such as *Eohippus*, had four front toes and three back toes, with the underside of the toes supported by a fleshy pad. This pad was still present, though reduced, in *Mesohippus*, which also differed in that its fore and hind feet had three toes each.

During the Miocene radiation of horses, however, there were different modifications to this three-toed arrangement. Horses lost the fleshy pads that supported their feet and stood upon a solid hoof, and the side toes were not necessarily in constant contact with the ground. Many horse lineages retained three toes for millions of years, although the reason for this is still not completely understood. (They might have been useful when horses were walking in mud, or even more in preventing overextension of the lower leg while running.)

Even among horses that were thought to have entirely lost their side toes, recent discoveries have shown that there was a messy period in which some individuals of a genus had three toes while others had one. This was the case with a population of a close relative of *Equus* called *Dinohippus* from the approximately eleven-million-year-old Ashfall Fossil Beds in Nebraska. An understanding of development solves the puzzle of how a population alternately presents archaic (side toes) and derived (no side toes) traits.

Horses are tetrapods, descendants of the first four-legged vertebrates that clambered about the Devonian mudbanks over 365 million years ago. When horses first evolved, they inherited the developmental pattern by which fingers are produced. In this developmental pattern the "pinky" side of the foot or hand develops first, sprouting off fingers in an arc toward the other side of the hand. The reduction of horse fingers and toes, then, was probably constrained by changes in development. Some rare living horses hint that this was the case.

For hundreds of years people have noted the birth of horses with extra toes. O.

FIGURE 85 – An "eight-footed Cuban horse," one of the rare polydactyl horses described by O. C. Marsh.

C. Marsh, in two papers published in 1879 and 1892, recorded several cases of horses that had their splint bones develop into lateral toes, complete with a "hooflet." Often there was just a single extra toe, on the inside of the leg. Through his work on fossil horses Marsh recognized this as an atavistic trait: the extra toes of the living horses were in the same place as the vestiges of the side toes in "normal" horses.

Yet at the time of Marsh's writing no one knew anything about DNA or the mechanism of inheritance. While the role of development in the evolution of the feet of horses has not yet been extensively studied, it is apparent that, even in living horses, the genetic triggers that cause the growth of lateral toes are still present, even if they are not often expressed. This would explain why some *Dinohippus* had side toes while others did not. At the time of the Ashfall *Dinohippus* specimens, the mutation that caused a change in the regulatory genes that controlled toe production had not yet spread to the whole population. Its spread might have been more of a case of genetic drift (a chance shift in the genetic makeup of a population) than an adaptive reduction of the toes. In other words, the horses with the mutation that constrained the development of toes might have been, by chance, more reproductively successful. *Dinohippus* with lateral toes were not inferior to the single-toed forms, so the disappearance of their type cannot be considered as any kind of "improvement."

In the opening of his 1891 summary called *The Horse*, the British anatomist William Henry Flower offered naturalists "insight into some of the fundamental principles of biology." The horse, familiar to all, could be used as a starting point from which to launch into almost any topic of inquiry about biology, especially evolution:

> The anatomy and history of the horse are, moreover, often taken as affording a test case of the value of the theory of evolution, or, at all events, of the doctrine that animal forms have been transmuted or modified one from another with the advance of time, whether, as extreme evolutionists hold, by a spontaneous or inherent evolving or unrolling process, or, as many others are disposed to think, by some mysterious and supernatural guidance along certain definite lines of change.

But while Flower was right about the utility of the horse in understanding evolutionary concepts, our growing understanding of horse evolution has failed to tease out any great unidirectional trend, supernaturally guided or not. The overall history of horse evolution has a stop-and-go pattern shaped by drastically changing environments and evolutionary constraints. Despite the utility of the animal to the rise of modern civilizations living horses, are just the last remaining twig of a previously much richer family.

Through the Looking Glass

"Whereas in truth Man is part a Brute, part an Angel; and is that
Link in the Creation, that joyns them both together."
—EDWARD TYSON, *The Anatomy of a Pygmy*, 1699

"In the first place it should always be borne in mind what sort of
intermediate forms must, on my theory, have formerly existed. I
have found it difficult, when looking at any two species, to avoid
picturing to myself, forms *directly* intermediate between them. But
this is a wholly false view; we should always look for forms intermediate between each species and a common but unknown progenitor; and the progenitor will generally have differed in some
respects from all its modified descendants."
—CHARLES DARWIN, *On the Origin of Species
by Means of Natural Selection*, 1859

In 1655, the French Calvinist Isaac La Peyrère gained the dubious distinction of publishing a book that was scorned by almost all who read
it. Titled *Prae-Adamitae* ("Pre-Adamites"), it was a heretical tome in
which Peyrère dared to suggest that Adam and Eve were not the first
and only people created by God. Instead there had been an early,
lawless world inhabited by Gentiles into which Adam, founder of the
Jewish people, had later been placed. Theologians and other scholars
were horrified. The Paris Parliament ordered the book burned, and
while Peyrère was traveling elsewhere on the European continent he
was tossed into jail for six months. The shocked public and clergy
demanded a recantation to quell their outrage. To atone for this sin,
Peyrère visited Pope Alexander VII in Rome, and blamed his Calvinist
upbringing for his wicked thoughts, though he held on to his belief in
pre-Adamites privately.

Peyrère's book was only a more fully formulated explanation of an
idea that had been raised many times before. It was therefore an easy tar-

get for attack, and numerous books were written to refute it. Yet Peyrère had no intention of undermining Scripture; he was seeking to assist the spread of Christianity.

Peyrère was led to his conclusions about the pre-Adamites by his reading of the Bible and the study of other cultures, in which he had found creation stories that differed from the two accounts presented in Genesis, and wondered whether they might indicate the existence of people—the pre-Adamites—prior to the ancestors of the Jews. This would explain some long-standing enigmas within Genesis. Who did Cain fear after being exiled for murdering his brother Abel? Where did Cain find his wife? Who peopled the city Cain built? Peyrère further supported his case for the pre-Adamites with a handful of verses from Paul's Letter to the Romans.

Many were unhappy with Peyrère's attempt to bridge this historical gap, but there was another void in nature that required an explanation. There could be no doubt that God was fond of a full creation, and therefore every organism could be ranked according to a Great Chain of Being. Frustratingly, there were some missing links in this scheme, and perhaps the most significant was the one between our kind and the rest of the animal creation.

For a time the "monstrous" races filled this role. Among the plethora of frightful, humanoid creatures thought to inhabit the far reaches of the globe were the Cynocephali, creatures with human bodies and doglike heads, and the Blemmyae, headless beings that wore their faces on their chests. These rude, half-human monsters lacked the one true hallmark of humanity, the soul.

By the time of Peyrère's heresy, however, physical proof of the monstrous races had failed to turn up. Monkeys, which had become more familiar to Europeans, were still the closest animal approximation to humans. As disturbingly humanlike as they could be (they were seen as walking warnings of the dangers of sin and vice), though, they were far too "low" to represent the step just below our kind.

In 1698, a sick, humanlike creature was brought to England. It had fallen ill from an infected wound sustained in transit from Angola, and soon after it arrived in London it perished. Such a rare specimen would not go to waste. The physician Edward Tyson, who had observed the animal shortly before its death, was awarded the precious and macabre opportunity of dissecting the "wild man." He undertook the task with great care, and the conclusions of his anatomical investigation were published the following year in his monograph *Orang-Outang, sive* Homo sylvestris: *or the Anatomy of a Pygmie*

FIGURE 86 – Tyson's "pygmie," most likely a juvenile chimpanzee. While Tyson thought it was a step below humanity in the created order, he gave it a walking stick as he believed the sick individual he observed would normally stand upright.

Compared with that of a Monkey, an Ape, and a Man.

Tyson called his subject an "Orang-Outang" on the basis of observations made more than five decades earlier by the Dutch physician Nicolaes Tulp. While Tulp was strolling through the menagerie of the Duke of Orange in 1641 he had spotted a hairy humanoid creature that seemed to fit the description of a creature rumored to live in Indonesia called the "Orang-Outang." Similar accounts of "wild men" had come from the region in which Tyson's "pygmie" was found. While held captive in Angola around the turn of the seventeenth century, the Englishman Andrew Battell reported that the local people were afraid of two humanlike creatures: the Engeco and the Pongo. Based upon folklore, Battell described the latter in vivid detail, painting the Pongoes as "hollow eyed" giants that slept in the trees and could even "build shelters from the raine." Even though they only ate fruit, their strength and ferocity were widely known, and the local people would not dare venture out alone if it was suspected that the Pongoes were about.

The animal Tyson expertly flayed on his dissection table was a poor fit for a Pongo, but he speculated that the tales of satyrs, monstrous beings, and "wild men of the woods" told of encounters with creatures like his "pygmie." Tyson concluded that "Our Pygmie is no man, nor yet the common ape; but a sort of animal between both."

Just a few decades after Tyson published his study on the "Pygmie," the eighteenth-century naturalist Carolus Linnaeus undertook the task of scientifically illuminating the divine pattern in nature. His classification system for organisms was first explicated in the 1735 book *Systema naturae*, but Linnaeus continued to produce new editions as he learned more. (The thirteenth and last edition was published in 1770.) Of prime importance, of course, was our position in relation to the rest of nature, and in the definitive tenth edition published in 1758 Linnaeus christened our species *Homo sapiens* and situated us among the "Orang-Outang," monkeys, and bats within the group Primates, which he regarded as the highest ranking of all organisms. [72] There was no question that we were the crowning glory of creation.

Critics reminded Linnaeus that our species was the only one created in the *imago Dei*, the image of God, and our possession of a soul cleaved us from the beasts. In 1747, he responded to such criticism from the German naturalist Johan Georg Gmelin:

> It matters little to me what names we use; but I demand of you, and of the whole world, that you show me a generic character, one that is according to generally accepted principles of classification, by which to distinguish between man and ape . . . I myself most assuredly know of none. I wish somebody would indicate one to me. But, if I had called man an ape, or vice versa, I should have fallen under the ban of all the ecclesiastics. It may be that as a naturalist I ought to have done so.

This was the most controversial part of Linnaeus's work. He had provided naturalists with a better method for naming organisms by giving each distinct kind of organism a genus and species name, but the proper hierarchical arrangement of those species was highly contested. Indeed, the proper place of *Homo sapiens* was made even more complicated by debates over whether different races of people were truly different species and how they might be ranked within the Great Chain of Being. Many European scholars considered the Greeks and Romans of classical antiquity to represent the pinnacle of the human type, and the racist culture of the time led them to suggest that the dark-skinned people of the tropics, who lived in the same geographic range as the apes, were closer to apes than people.

By the early nineteenth century, however, the Great Chain had nearly rusted through. Nature was not only full, but overflowing with a variety of forms that were impossible to rank without resorting to subjective criteria. How could the diversity of the stunning Indonesian birds of paradise be ranked from higher to lower? How could the varieties of dogs be graded from superior to inferior?

Naturalists were also beginning to question nature's immutability at this time. Did God have such a fondness for beetles that he handcrafted each and every species, or was there some kind of natural law to control the beetle manufacturing process without further divine intervention? Evolution was still treated as a heretical idea, but the science of physics had made it acceptable to think that God had ordained natural laws to govern the workings of the universe.[73]

Perhaps the production of new species was regulated by such a law, but even as proto-evolutionists toyed with this idea many stolidly affirmed that humans were a special case. To think that we had evolved from some "lower" creature, as Lamarck suggested, was revolting. Humanity clearly contained the spark of the divine, and this position was tenaciously defended by Richard Owen. In 1857 he placed our species, and our species alone, in a new subclass of mammals he called the "*Archencephala*." Our upright posture and our "extraordinarily developed brain" justified this move, Owen argued, but more important was the position at the head of nature given to us by God. "It is He that hath made us; not we ourselves," Owen reminded his audience, and he admonished his listeners that "This [human] frame is a temporary trust, for the uses of which we are responsible to the Maker."

As Linnaeus had pointed out in the previous century, however, any division between human and ape would have to be based on solid science. Appeals to aesthetics and special pleading were inadmissible. If cherished characteristics such as speech, thought, and spirituality truly made us distinct, their anatomical correlates would have to be identified. Owen believed he had found these in several minute features of the brain, including part of the temporal lobe called the hippocampus minor. Apes did not have these features, thus explaining why they were silent. Despite there being no anatomical reason that apes could not speak, Owen announced that chimpanzees, the "Ourang-Outangs" of Tulp and Tyson, had no language because they were cerebrally deficient.[74] Science had finally drawn a bright line between human and ape.

Charles Darwin balked at Owen's assertions. In a letter to a friend, the botanist Joseph Hooker, Darwin remarked, "Owen's is a grand Paper; but I cannot swallow Man making a division as distinct from a Chimpanzee, as an ornithorhynchus [duck-billed platypus] from a Horse: I wonder what a Chimpanzee wd. say to this?" Yet Darwin refrained from speaking on behalf of chimpanzees. His evolutionary mechanism was controversial enough without emphasizing our ape ancestry, but this did not fool anyone. The implications of the Darwin–Wallace mechanism were clear: not only had we evolved, but living apes approximated what our own ancestors might have looked like.

Owen remained steadfast in his conclusions. To him humans had been specially set apart from all other creatures, and he continued to insert that we were separated from apes by the hippocampus minor. Some intellectual allies of Darwin mobilized to combat the claim, and leading the counterattack was T. H. Huxley. In the January 1861 issue of the

Natural History Review, Huxley demonstrated that the brain structures Owen claimed to be uniquely human were also present in apes, and the anatomists John Marshall and William Henry Flower came to the same conclusion in their research. But Owen refused to back down. While such structures were present in apes, Owen argued, they were not exactly the same as their homologous parts in our brains and therefore did not merit the same anatomical name. Owen was arguing over the definition of these parts, while Huxley and his allies were primarily concerned with their presence, making the academic conflict intractable.

This debate over a bit of brain matter was of high importance. Questions about our place in nature after Darwin's evolutionary mechanism was unleashed often relied on determining what made us different from our vulgar ape ancestors. Even though Darwin and other naturalists made it clear that we did not evolve from any living species of ape, the known array of orangutans, gorillas, and chimpanzees were the only available window into our ancient past. Paleontologists had found a handful of fossil apes, such as the gibbonlike *Dryopithecus* from France, but no transitional "ape-men" had been found. In fact, in 1859 scientists had only just begun to understand how far back our ancestry could be drawn.

Until the mid-1800s it had been believed that our species had been in existence scarcely longer than the span of recorded history. This was despite the fact that evidence of our antiquity had long been available. Stone "thunderbolts," or *ceraunia*, had been found in Europe for centuries. They were often considered to be mineralogical curiosities, like fossils or rocks shaped like human body parts, and were ascribed supernatural powers. The sixteenth-century Italian scholar Michael Mercati disagreed:

> Most men believe that *ceraunia* are produced by lightning. Those who study history consider that they have been broken off from very hard flints by a violent blow, in the days before iron was employed for the follies of war. For the earliest men had only splinters of flint for knives.

This complemented the logic of the Roman philosopher Lucretius, who over 1,500 years earlier supposed that warfare had begun with stone weapons. Just how old these implements were, however, was open to interpretation. In 1679, an English pharmacist named John Conyers found stone tools near the skeleton of an elephant. The presence of the elephant was as mysterious as the tools, and when the find was announced after Conyers's death the collector of antiquities John Bagford proposed that it had been brought to England by the Roman emperor Claudius in the first century AD. The tools must have been of the same age, and the fact that native peoples in North America and elsewhere made similar implements showed that they must have been made during the span of recorded history.

When John Frere discovered similar artifacts in Suffolk in 1797 he came to a different conclusion. He thought that the tools could have represented the work of people from "a very remote period indeed, even beyond that of the present world." His peers in the Society of Antiquaries were skeptical, however, perhaps because much other evidence of "fossil humans" did not stand up to scrutiny. The French anatomist Georges Cuvier, for instance, had been presented with elephant bones, giant salamander skeletons, and other miscellaneous fossils as bona fide human remains by amateur naturalists. Even when a true human skeleton said to be of great age was presented to Cuvier in 1823 he found no conclusive sign to that effect.

That same year William Buckland delivered a lecture on a partial human skeleton he had excavated from Goat Hole on the Welsh Coast nicknamed the "Red Lady of Paviland." At first Buckland thought that the bones were of a man who had been attacked by smugglers, but he soon changed his mind. The skeleton was stained red with ochre and found among shells and carved artifacts, which Buckland thought to be the hallmarks of a witch (though he did not state this interpretation publicly). He reasoned that she had lived in the cave around the time of the Roman occupation of England. After his lecture on the "Red Lady" he found a scrap of paper on the classroom floor with a poem written on it:

> Have ye heard of the woman so long underground?
> Have ye heard of the woman that Buckland has found,
> With her bones of empyreal hue?
> O fair ones of modern days, hang down your heads,
> The antediluvians rouge`d when dead,
> Only granted in lifetime to you

Buckland was amused by the verses, but he could not agree that the Red Lady was "antediluvian." Even though the skeleton was found alongside the fossils of extinct mammals, at the time Buckland thought that the mammal bones had been washed into the cave during the global flood. The Red Lady did not occupy the cave until much later, sometime after the repopulation of the world.[75]

Still, the Bible was clear that antediluvian humans were a historical reality. Where were their remains? Buckland supposed that they had lived in Asia, a land that had been prepared by God for human habitation. Europe, by contrast, would have been inhospitable to humans, as it had been home to big cats, elephants, hippos, rhinoceros, and hyenas while devoid of domestic animals that would have been useful to humans. This point was later articulated by Richard Owen in his 1846 book *A History of British Fossil Mammals, and Birds*:

> When we are informed that, in some districts of India, entire villages have been depopulated by the destructive incursions of a single species of large Feline animal, the Tiger, it is hardly conceivable that Man, in an early and rude condition of society, could have resisted the attacks of the more formidable Tiger, Bear, and Machairodus [saber-toothed cat] of the cave epoch. And this consideration may lead us the more readily to receive the negative evidence of the absence of well-authenticated human fossil remains, and to conclude that Man did not exist in the land which was ravaged simultaneously by three such formidable Carnivora, aided in their work of destruction by troops of savage Hyacnas.

The complexities of cave geology threw further doubt on the proposition that these animals and humans had been coeval. Unraveling the order in which cave strata were deposited was a difficult task, and it was often simpler to assume that stone tools and human bones were recent contaminants that had become mixed with older fossils. The activities of amateur geologists further confounded efforts to understand the confusing conglomerations of bones. While the sciences of archaeology, geology, and paleontology were born from the inquisitiveness of amateurs, by the 1840s these sciences were becoming professionalized. When amateurs plucked tools and human bones from the ground without taking copious notes on the excavation it was all the easier for professionals to dismiss discoveries and insist that the sudden appearance of humans marked a new, distinct geological age.

The growing unease between professional and amateur scientists was on full display during the excavation of a particularly important cave in southwest England. While out to quarry limestone in early 1858, the entrepreneur John Philp stumbled onto a cache of fossils in Brixham cave in Devonshire. He was sure that the strange bones could turn a profit, so he set up a little museum inside the cave, and word of the *in situ* exhibit grabbed the attention of amateur geologist William Pengelly. When Pengelly inquired if he and his fellows in the nearby Tourquay Natural History Society could study the cave, Philp was happy to oblige, but only if the price was right.

Unfortunately, the Tourquay Natural History Society had little hope of making the rent Philp required, but their luck changed. In April of 1858, Pengelly told the fossil mammal expert Hugh Falconer about the cave, and Falconer agreed that it had potential. This was a chance to study a fossil-bearing cave in detail. Falconer promised to mention the situation to his fellows at the Geological Society of London.

Upon hearing Falconer's report the Geological Society agreed that the cave should be studied, and they set about securing funding. They also formed a cave committee consisting of Pengelly, Falconer, Charles Lyell, Richard Owen, Andrew Ramsay, and Joseph Prestwich. By May, excavations were ready to begin. The project was carried out by two teams. Members of the Tourquay Natural History Society supervised the digging carried out by hired help, and the fossils were sent back to the professional geologists in London for analysis. This was a precise enterprise, and Pengelly made careful geological notes as the strata were peeled back and the fossils removed. By August over 1,500 mammal bones were recovered, but the workers also turned up a few relics that were even more intriguing.

Before the summer ended, seven stone tools were found mingled with the remains of extinct cave bears, hyenas, and rhinoceros. At long last the elusive proof of "men among the mammoths" had been found, but most of the Geological Society experts were skeptical. Joseph Prestwich, in particular, worried that if the find was reported too soon religious outrage would overwhelm whatever scientific value the tools held. These fears put Falconer's enthusiastic report on the artifacts through a month-long process of editing to make it less controversial.

The scientific censors overshot their mark. Despite the buzz surrounding Falconer's report, it failed to elicit much of a response when it was finally read before a packed audience of London scientists. The evidence was too flimsy to push back the emergence of humans, and the fact that most London geologists had not even seen the tools made it easy to downplay their significance. Yet the excitement over the tools was not so much over their manufacture but their age, a point on which Pengelly's notes were essential. This strained the relationship between Falconer and Pengelly, as the former began to lecture Pengelly on proper geological note-taking.

Tensions mounted over the unequal division of labor between the Torquay amateurs and the London theorists, too. This came to a head when Torquay Natural History Society member Edward Vivian published an article that attempted to reconcile Brixham cave with a literal reading of Genesis. The London academics were shocked and embarrassed. They ordered a halt to any publications about the cave, and even though it had previously been understood that the Brixham fossils would be returned to Torquay, the experts now insisted that the bones and artifacts stay in London.

As the London–Torquay association was suffering, so was Falconer's health. He set off on a salutary trip to the Mediterranean intending to investigate a few interesting fossil sites along the way. One of his first stops was in Abbeville, France, the home of the controversial scientist Jacques Boucher de Perthes. During the 1830s the French naturalist Paul Tournal discovered stone tools alongside bones of extinct mammals, and Perthes made similar finds around Abbeville in 1838. He presented these finds in the first volume of his 1846 work *Antiquités celtiques et antédiluviennes* and boldly proclaimed that the "rude stones" mixed with the extinct fauna "prove[d] the existence of Man as surely as a whole Louvre would have done." This assertion led to a barrage of jeering opposition and scientific criticism from his peers. Falconer, too, was skeptical of Perthes' claims, but he recognized that a few of the French tools resembled those from Brixham.

Falconer wrote to Prestwich to suggest that further investigations of Abbeville might be worthwhile, and Prestwich soon visited the nearby gravel pits with his friend John Evans. The duo left Abbeville empty-handed, but as soon as they returned home to England Prestwich received a message that a stone tool had been found and left *in situ* for him to examine. Prestwich and Evans dashed back across the English Channel and found precisely what they had been hoping for.

Whereas the age of the Brixham cave was difficult to determine, the stratigraphy of the French site in the Somme Valley was better known. As a result, Prestwich was able to convincingly show that the tool had been deposited at the same time as the extinct mammal fossils around it. Even so, Prestwich was cautious. He did not want to move the age of humanity backward in time any more than he wanted to move ancient mammals forward. He did note that the appearance of humans did not mark the beginning of a distinct geological period; our forebears had once inhabited a more ancient, unfamiliar, and dangerous world.

Prestwich encouraged his colleagues to visit the tool-bearing sites to see for themselves, and the geological community quickly reached a consensus. Fresh from the Somme Valley sites, Charles Lyell publicly announced the coexistence of human and ancient mammals before the 1859 meeting of the British Association for the Advancement of Science. "No subject has lately excited more curiosity and general interest among geologists and the public than the question of the antiquity of the human race," Lyell said, and there was now compelling evidence that we had a more ancient origin than had been previously thought. It is no surprise that Lyell referred to *On the Origin of Species* (published that November) in the same speech, for the time had now come to more carefully examine the ancient history of our species.

This new understanding stirred up opposition, but eventually even critics like Thomas Wright and John Henslow were won over by the evidence. A new consensus had emerged almost overnight, though geologists still faced an old dilemma. Where were the bodies of the tool-makers? Human fossils that had been previously found (such as the Red Lady) were too poorly documented to be used as evidence. Outside of their geological context no one could be sure how old the fossils were.

Similar problems plagued the study of human fossils found in a Neander Valley limestone quarry near Düsseldorf, Germany. It was there that in 1856 laborers found a skull cap and portions of the limbs, ribs, hips, and shoulders of what they believed to be a cave bear. They presented the fossils to the local naturalist Johann Carl Fuhlrott, but he came to a different conclusion: the bones were human. Excited by this prospect, Fuhlrott brought the bones to the attention of the Bonn anatomist Hermann Schaffhausen, who agreed with Fuhlrott's interpretation. They announced the discovery jointly in 1857 as the remains of a pre-modern human "of a barbarous and savage race."[76] The announcement was controversial from the start.

Like other human fossils, the Neander Valley bones had been exhumed by non-scientists, and no fossils useful for determining their age were found alongside them. Even more controversial was the question of what kind of human the skeleton represented. Although of a comparable stature to our species, the bones were more robust and the skull cap was low-domed with a heavy brow ridge. Was it one of our ancestors, an ancient "cousin," or just an aberrant *Homo sapiens*?

Numerous hypotheses were proposed during the following years. August Franz Mayer, a peer of Schaffhausen at Bonn, thought that the Neander Valley skeleton belonged to a Mongolian Cossack that had been driven into Ger-

FIGURE 87 – The first recognized Neanderthal skull as seen from the side, front, and top.

many during the Napoleonic Wars, the slouched posture and bowed leg-bones having been effected by a life spent on horseback; Pathologist Rudolf Virchow, by contrast, deemed the specimen a *Homo sapiens* that had been distorted by disease; and their colleague Pruner Bey proposed that the skeleton was a dead ringer for "a powerfully organized Celt, somewhat resembling the skull of a modern Irishman with low mental organization." Even T. H. Huxley thought that the bones fell within the range of variation of our species and at best represented a primitive throwback:

> In no sense, then, can the Neanderthal bones be regarded as the remains of a human being intermediate between Men and Apes. At most, they demonstrate the existence of a Man whose skull may be said to revert somewhat towards the pithecoid [ape] type—just as a Carrier, or a Pouter, or a Tumbler [breeds of pigeons], may sometimes put on the plumage of its primitive stock, the *Columba livia*.

The conclusion of the Anglo-Irish geologist William King, however, was closer to that of Fuhlrott and Schaffhausen. He thought that the bones belonged to a different species of human, which he named *Homo neanderthalensis* in 1864. (American paleontologist E. D. Cope, presumably unaware of King's work, proposed the same name for the species after more remains were recovered from Spy, Belgium.) That the Neanderthals were different was apparent when they were compared to our own Stone Age precursors, the Cro Magnons, who had been described in 1868 by the French geologist Louis Lartet. Still, the Neanderthals were too similar to us to shed much light on our ape ancestry. True fossil "ape-men" that filled the chasm between us and our earlier primate ancestors had yet to be found.

The dearth of transitional human fossils must have frustrated Darwin. He certainly could have used them when he wrote his 1871 book devoted to human evolution, *The Descent of Man, and Selection in Relation to Sex*. As things were, Darwin could only restate the fortuitous nature of the fossil record and assure his readers that the place where the bones of our ancestors were most likely to be found had not yet been searched. Appealing to the "law of fossil succession" he had identified as a youth in South America, Darwin proposed that Africa might yield the remains of the earliest humans.

> In each great region of the world the living mammals are closely related to the extinct species of the same region. It is therefore probable that Africa was formerly inhabited by extinct apes closely allied to the gorilla and chim-

panzee; and as these two species are now man's nearest allies, it is somewhat more probable that our early progenitors lived on the African Continent than elsewhere.

But this was anything but a certainty. Just below this famous passage Darwin remarked:

> But it is useless to speculate on this subject, for an ape nearly as large as a man, namely, the *Dryopithecus* of Lartet, which was closely allied to the anthropomorphous *Hylobates* [gibbons], existed in Europe during the Upper Miocene period; and since so remote a period the earth has certainly undergone many great revolutions, and there has been ample time for migration on the largest scale.

Nor did Darwin believe that we shared a recent common ancestor with chimpanzees or gorillas. Of all the apes they were the most similar to us on a superficial level, but in an 1868 sketch of the human family tree Darwin drew in a notebook, he placed a deep divide between them and us. Chimpanzees, gorillas, orangutans, and gibbons all shared a relatively recent common ancestor, but humans had arisen from some earlier ape stock before the later branches diverged. Humans were on their own divergent branch that had split off early, or in Darwin's words, it seemed "that some ancient member of the anthropomorphous sub-group gave birth to man."

The young Dutch anatomist Eugène Dubois thought he had a good idea of where to look for the fossils of such creatures. For him the Asian orangutans and gibbons were much more like us than the African apes, and the discovery of a fossil ape from the Siwalik Hills of India reinforced his hunch that our origins lay in the East. The Indonesian island of Sumatra contained limestone deposits of about the same age as the Indian site, making it a promising place to look for human ancestors. To get there Dubois enlisted as a medical officer in the Dutch army. In 1887, he left home, family in tow, to begin his search.

Soon after he arrived at Sumatra, Dubois found enough fossil mammals to convince the Dutch government that his expedition was worthwhile. They provided him with two engineers and fifty forced laborers to continue his work, but one engineer proved incompetent, the other fell ill, and many of the laborers deserted. This was just as well. The fossils the crew exhumed were too recent, and there was little hope of finding an "ape-man" among them. In 1890, to start over again Dubois transferred to Java where he put a new crew to work in the vicinity of

FIGURE 88 – A photograph of the *Pithecanthropus* skullcap and the associated femur and tooth.

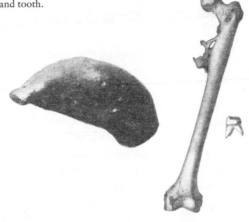

the Solo River. Piece by piece, parts of a human-like skeleton were culled from the sediment: a jaw in 1890, a molar and skullcap in 1891, and a left femur in 1892.

Dubois called his find *Anthropithecus erectus*, or the "erect man-ape." This was a fitting name for a creature with a femur fit for upright walking and a brain size about 1.75 times that of a chimpanzee, but Dubois had made an error in his calculations. In December of 1892, Dubois discovered his mistake and realized that the brain of *Anthropithecus* was 200 cubic centimeters larger than he had first surmised. This placed it closer to humans than to apes, and Dubois did a little taxonomic rearranging to rechristen his find *Pithecanthropus erectus*, the "erect ape-man." (Ernst Haeckel had informally proposed a hypothetical human ancestor he called *Pithecanthropus alaulus*, or "speechless ape-man," some years earlier.) Excited by his find Dubois began sending reports of his discovery to colleagues back home.

When Dubois returned to Europe in 1894 he was hit by a wave of controversy. Experts disagreed on what type of creature *Pithecanthropus* was, if indeed the bones represented just one creature. The femur was very humanlike and clearly from a bipedal animal, but the low-domed skull was more apelike. This led some experts to argue that the bones of a human and an ape had become mixed together, but Dubois remained firm in his interpretation:

> The more I myself have studied these fragments the more firmly I have been convinced of this unity of origin; and at the same time it has become ever clearer to me that they are really parts of a form intermediate between men and apes, which was the ancestral stock from which man was derived.

Even though Dubois received accolades for his achievements, the identity of "Java Man" remained contentious. The refusal of some of his peers to agree with him frustrated Dubois, and he was aghast when an unauthorized description of *Pithecanthropus* was made by the German anatomist Gustav Schwalbe in 1899. It seemed that the argument over his discovery was slipping out of his hands. Without warning Dubois closed off all access to the bones in 1900, but he had a plan that would ultimately establish *Pithecanthropus* in its rightful place between human and ape.[77]

Using the femur and skullcap as guides, Dubois wrangled with the relationship between brain and body size. Based upon the femur, it appeared that *Pithecanthropus* would have closely matched our species in stature, but the calculated brain size was only about two-thirds of our own. This did not match Dubois' expectations. Dubois believed that there was an internal driving mechanism in evolution that caused brain size to double in respect to body size in each advancing stage of evolution. If *Pithecanthropus* had a body like ours, this ratio would be thrown off, but Dubois reasoned that if it had the proportions of a large gibbon then it would fall halfway between human and ape.

Dubois took his time with *Pithecanthropus*, but after more than two decades of secrecy his colleagues had grown impatient with him. It was unfair that important fossils collected during a government-sponsored expedition were held under lock and key. Under threats and formal complaints, Dubois relented and restored access to the fossils in 1923. Even though many of his peers did not accept his new conclusion that *Pithecanthropus* looked like a giant gibbon, and it could still not be determined whether it was truly one of our ancestors, Java Man was generally welcomed as an early fossil human. It was the first of the fossil ape-men to be scientifically described, but it was no longer the only one.

In 1908, the English archaeologist Charles Dawson was given some scraps of bone found in a gravel pit in Uckfield, East Sussex. A collector of stone tools and other artifacts, Dawson scoured the site for more bone fragments, and eventually he showed what he had found to geologist Arthur Smith Woodward. In September 1912, Woodward and Dawson recovered even more bone fragments from the gravel pit. When they were all put together the pieces represented a partial skull

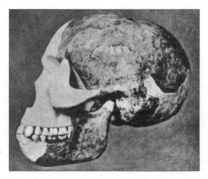

FIGURE 89 – The reconstructed skull of "Piltdown Man."

and the lower jaw of an ancient human they decided to call *Eoanthropus dawsoni*, popularly known as "Piltdown Man."

The reconstructed skull of *Eoanthropus* showed a strange combination of features. The lower jaw was similar to that of a juvenile ape, but the skull was reminiscent of *Homo sapiens*. Indeed, the high-domed head confirmed that an increase in brain size led the way in human evolution, for this creature had evolved a large brain before all the ape traits were lost. The French anthropologist Marcellin Boule thought that the jaw seemed so "ape-like" because it was actually from an ape, and his American colleague G. S. Miller came to a similar conclusion.

These concerns echoed the *Pithecanthropus* debate from decades before, and for much the same reasons. The fragmentary nature of the bones and the span of time it took to recover them cast doubt on whether all the pieces belonged to only one individual. It was an osteological jigsaw puzzle that led Miller to write, tongue-in-cheek, "*deliberate malice could hardly have been more successful than the hazards of deposition in so breaking the fossils as to give free scope to individual judgment in fitting the parts together.*"

Even more perplexing was that *Eoanthropus* was in the wrong place to be a human ancestor. Latent racism made many anthropologists balk at the notion that our species had evolved in Africa, and many scientists believed that Asia was a more suitable place for our kind to have emerged. The American Museum of Natural History launched several expeditions to Mongolia during the 1920s to dig up evidence for this hypothesis, but all they found were ancient mammals and dinosaurs. A different team working in China had better luck.

In 1921, the Austrian paleontologist Otto Zdanksy, working with Swedish geologist John Gunnar Andersson, American paleontologist Walter Granger, and Canadian anatomist Davidson Black, discovered a humanlike tooth at Dragon Bone Hill in Zhoukoudian, China. It was not much to go on, but Black wrote a short communication to *Nature* suggesting that "man or a very closely related anthropoid [ape] did exist in eastern Asia" at the same time that *Pithecanthropus* lived in Indonesia. This tooth was matched by two others found in 1926, and Black used them to establish a new kind of fossil human at the site, *Sinanthropus pekinensis*, even though the rest of the skeleton proved elusive.

Three more years passed without further sign of *Sinanthropus*, but in November 1929 the team found a 130-foot-deep hollow they dubbed "Ape-Man Cave." The onset of harsh winter weather forced many of the experts to end their field season, but Chinese scientist Pei Wenzhong and a few workers stayed on to fight the frozen ground for fossils. On

December 2, their efforts were rewarded with the first "Peking Man" skullcap. It would not be the last. By 1937, numerous *Sinanthropus* bones had been recovered, including six low-domed and heavy-browed skulls. Some scientists, like Franz Weidenreich, thought the skeletons represented a type of human more primitive than *Pithecanthropus*, but more discoveries were needed to determine the exact relationship between the two.

Research halted when the Japanese army began its occupation of China. The looming prospect of war threatened the safety of the bones, and it was decided that the *Sinanthropus* fossils would be sent to the United States for safekeeping. In early December of 1941, the fragile cargo was readied for its journey and placed in the care of a group of American marines—but on December 7 war broke out between the United States and Japan when Japanese forces attacked the American naval base at Pearl Harbor, Hawaii. Military forces on both sides sprung into action, and the American soldiers charged with protecting the *Sinanthropus* bones were captured before they could leave China. What happened to their ancient cargo is unknown, and the precious collection of bones was never seen again.[78]

The loss of these relics was devastating, but by this time a neglected set of fossils from Africa would turn out to be even more important to understanding human origins. While teaching anatomy at the University of the Witwaters, and in Johannesburg, South Africa, in 1924, the Australian-born anatomist Raymond Dart told his students to keep a sharp eye out for interesting fossils. When his student Josephine Salmons told him she had spotted a fossil baboon skull in the living room of the director of the Northern Lime Company, Dart wondered if the nearby lime quarry might contain other rare fossils, and he arranged to have two crates worth of bone-bearing limestone that had already been excavated delivered to his home.

When the crates arrived, Dart was playing host to a friend's wedding in which he was also filling the role of best man. He was supposed to be getting dressed, but the arrival of the fossils was just too exciting. He decided to at least have a peek before carrying out his duties. The first crate contained little of interest, just a few fossil turtles and miscellaneous bone fragments, but he was shocked by what he saw when he opened the second. When he took off the lid he saw the fossilized cast of a primate brain, and a large one at that, with the rest of the skull present in another chunk of rock. Dart was astonished; no one had ever found such a thing before.

With the groom plaintively tugging at this sleeve, Dart returned the fossils to their packaging and performed his ceremonial duties, but for him the wedding could not have ended soon enough. As soon as the last of the guests left he went back to have another look at the fossil. It was an extraordinary specimen, but Dart had few resources to help him prepare his treasure. He whittled away at the limestone with a hammer, chisel, and one of his wife's knitting needles, and on December 23 the visage of the primate came into view.

The ancient face was flatter than expected for an ape, and it lacked the heavy brow ridges of the Neanderthals and *Pithecanthropus*. More surprising, though, was that the foramen magnum, the hole in the bottom of the skull through which the spinal cord exits, was oriented downward. In quadrupedal animals, like dogs, the foramen magnum is oriented backward, while in our species it is situated underneath so that the skull sits on top of the spinal column. The skull Dart was examining had an opening oriented further forward than that seen in chimpanzees, and this suggested that it had been a bipedal animal. Combined with the exceptionally large brain size, these features linked his fossil closely to our ancestry. Dart rushed a manuscript to *Nature* in which he named the fossil *Australopithecus africanus*, the "southern ape from Africa."

Other physical anthropologists were not very impressed. Like Tyson's "Pygmie," juvenile apes were known to show similarities to human skeletons that disappeared as the apes aged. Many anthropologists felt that Dart had simply found a fossil ape that was of little relevance for human evolution. The general consensus of anthropologists was made clear when Dart brought his "Taung child" (named for the quarry from which it was found) to London to show the scientific elite. Following a well-received presentation on the discoveries at Dragon Bone Hill illustrated with lantern slides, Dart simply stood in front of his audience and spoke about the miniscule skull in his hand. He felt overwhelmed and underprepared, and he had failed to change the opinion of many of his colleagues that *Australopithecus* was a backwater genus of ape made irrelevant by the exciting work being done in Asia.[79]

Yet Dart had a strong ally in the Scottish paleontologist Robert Broom. While Dart returned to his specialty of neuroanatomy, Broom visited numerous South African caves and quarries in search of more *Australopithecus* fossils. He found them in abundance, including the skulls and brain casts of adults. This led Broom to name a number of new genera and species that were later absorbed into existing categories. Among the new fossils was a second, more robust type of australopithecine with massive jaws and crushing teeth he named

Paranthropus robustus in 1938. These were impressive finds, but many experts still thought these creatures were just aberrant apes. They were not ancestral to humans, many anthropologists agreed, and so were only of passing interest.

Other fossil apes seemed to have more potential to fit into our pedigree. On Rusinga Island in Kenya's Lake Victoria, the English anthropologist Arthur Hopwood found a fossil ape he named *Proconsul* in 1933. The next year, G. Edward Lewis announced another fossil ape, *Ramapithecus*, from the Siwalik Hills of India. These apes were old enough and generalized enough to possibly have given rise to the earliest humans, but there still remained a wide gap between them and *Homo erectus* (the new name for the "Sinanthropus" and "Pithecanthropus" fossils). Could the australopithecines of Dart and Broom have fit in the void between fossil human and fossil ape? Most of the pronouncements about them had been based upon casts and photographs, so when the Kenyan-born anthropologist Louis Leakey organized a conference on human prehistory in Nairobi in 1947, his English colleague W. E. Le Gros Clark took the opportunity to travel further afield to have a closer look at the "southern apes." [80]

Armed with copious notes on ape anatomy Le Gros Clark examined the original australopithecine fossils and visited the sites where they had been found. He was certain that australopithecines were nonhuman apes when he arrived, but the evidence he saw forced him to reject this conclusion. Now it was "hardly possible to overemphasize their significance," for their anatomy clearly showed that "there must be a real zoological relationship between the Australopithecinae and the Hominidae [humans]."

There were several features that strengthened this relationship between the australopithecines and humans. One important observation was that the end of the humerus (upper arm bone) of *Paranthropus* was not as flared and robust at the elbow joint as in chimpanzees. Instead, it resembled a human humerus and suggested that the australopithecines were probably not walking on their knuckles. This was confirmed by the orientation of a bone in the foot, the astragalus, which articulates with the leg to form the ankle joint. Its shape is a vital clue in determining posture, and the australopithecine astragalus more closely resembled the same bone in our skeleton than its counterpart in chimpanzees or gorillas. This matched the orientation of the end of the femur that formed the knee joint. It was directed inwards like ours— another trait associated with human bipedalism.

Combined with other similarities, these features allowed Le Gros

Clark to demonstrate that the australopithecines were bipedal creatures more closely related to us than living apes, and he deemed them the "little modified survivors of the ancestral stock from which, at a still earlier date, the line of human evolution originated." He wasted no time relaying these findings to the conference attendees in Kenya, and he reiterated his statements in print when he returned home to England. Many of his colleagues who had spurned the australopithecines, like Arthur Keith, changed their mind on the subject. Perhaps Africa held clues to our evolution, after all.

Le Gros Clark's conclusions signaled a turning point in paleoanthropology and it was underscored by a lecture the vertebrate paleontologist G. G. Simpson delivered at the Cold Spring Harbor symposium on human origins in 1950. Despite their interest in evolutionary questions, he said, paleoanthropologists had largely ignored the development of the modern evolutionary synthesis. Many were still cooking up almost fanciful stories and appealing to internally driven mechanisms of evolution that other paleontologists, Simpson foremost among them, had cast out. Simpson's criticisms cut deep, but they were essential to give paleoanthropologists a more solid evolutionary framework in which to interpret their finds.

The revitalization of paleoanthropology coupled with the intriguing discoveries made in Africa precipitated a conference on African hominids held in England in 1953. All the major extinct human fossils were discussed, save for one: Piltdown Man.

As discoveries were made elsewhere, the Piltdown fossils became more of an anomaly that did not fit in with the rest of the human evolutionary picture. This change perplexed one of the conference attendees, Joseph Weiner. Even though he found the idea repugnant, Weiner thought it possible that someone had planted the bones in the countryside gravel pits nearly half a century earlier. He reinvestigated what was known about the fossils, and the more he dug the more suspicious the evidence became. After expressing his concerns to colleagues W. E. Le Gros Clark and Kenneth Oakley, the three confirmed their fears: "Piltdown Man" was a hoax.[81]

As it turned out, Piltdown Man was a chimera. The jaw came from an ape and the skull was *Homo sapiens*, and the forgery extended beyond the "Eoanthropus" fossils alone. Nearly everything of scientific significance, from the stone tools to the bones of extinct mammals from the site, had been artificially manipulated and deposited.[82]

The Piltdown fiasco had given anthropology a black eye. No paleontologist had suspected that one of their own colleagues held such a capacity for deceit. Still, work carried on, and with the reputation of the australopithecines rehabilitated, Raymond Dart returned to paleoanthropology to paint a horrifying picture of our past. The depravity of our own species had been put on full display during World War II. For Dart, these were just echoes of the violent birth of humanity. Vast accumulations of battered bones in South African caves led Dart to think that australopithecines had an "osteodontokeratic" culture in which they used bones, teeth, and horns as tools to kill prey. Even more horrific was the prospect that the smashed skulls of some australopithecines represented the birth of our kind:

> The blood-bespattered, slaughter-gutted archives of human history from the earliest Sumerian records to the most recent atrocities of the Second World War accord with early universal cannibalism, with animal and human sacrificial practices or their substitutes in formalized religions and with the worldwide scalping, head-hunting, body-mutilating, and necrophilic practices of mankind in proclaiming this common bloodlust differentiator, this predaceous habit, this mark of Cain that separates man directly from his anthropoidal relatives and allies him directly with the deadliest of Carnivora.

The remains in the hollows were clues to ancient crime scenes where murderous man-apes clubbed each other to death and feasted on the flesh of the fallen. Blood flowed freely across the plains as australopithecines hunted not only game animals, but each other. Perhaps they even became the prey of later, more advanced hominids. *Homo habilis*, described in 1964 by Louis Leakey, Philip Tobias, and John Napier, was such a contender.

The story of *Homo habilis* actually began several years before, with the description of an entirely different kind of human. In the summer of 1959, Louis and Mary Leakey were searching Olduvai Gorge in Tanzania for early human fossils when Mary spotted the fragments of a skull weathering out of the hillside. They worked through August to collect all the pieces they could find, and as they picked over the site they found stone tools and the shattered bones of other mammals. For Louis, the implications of this association were clear. Whatever this human was, it had been a toolmaker that processed the carcasses of mammals for meat, and since it used tools it was most certainly ancestral to our own species.

Louis remained steadfast in this interpretation, even as the skull Mary pieced together did not look anything like one of our direct ancestors. The heavy brow ridge, low-domed skull, flaring cheek bones,

FIGURE 90 – The skull of *Paranthropus boisei* recovered by Louis and Mary Leakey at Olduvai Gorge

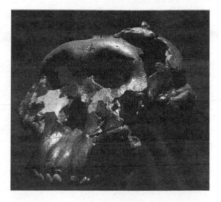

and enormous teeth identified the human as a species of *Paranthropus*, similar to the humans Robert Broom had found in South Africa, but Louis could not accept this. He had a habit of thinking of his fossils as true ancestors of *Homo sapiens* and those of everyone else as dead-end side branches less relevant to our ancestry. He asserted that there were subtle differences between his new fossil and *Paranthropus* that placed it closer to our ancestry, and he decided to name it *Zinjanthropus boisei* (though he informally called it the "Dear Boy").

This identification created some new problems. At the time of its discovery it was thought that the deposits from which the skull came were only 600,000 years old, which would make "Zinj" a creature of the ice age. (Repeated testing with absolute dating techniques would later show that the skull was about 1.75 million years old.) This was not very much time for something as distinct as Zinj to evolve into *Homo sapiens*, so Leakey had to come up with a way to speed up the evolutionary process to span the gap. The stone tools seemed to point at an adequate solution. In a lecture given to the South African Archaeological Society in 1960, he proposed that the use of stone tools had greatly accelerated the evolution of something like Zinj into us. We had domesticated ourselves in less than 400,000 years.

Leakey abandoned this view almost as quickly as he had proposed it. That same year, his son Richard and his wife, Mary, began to find bones from a prehistoric human that was anatomically more similar to us than Zinj was. Louis quickly became convinced that these fossils were the remains of the true toolmakers, perhaps the earliest members of *Homo* yet found, and he kicked Zinj off to a sidebranch. (It was later grouped with the robust australopithecines as *Paranthropus boisei*.) Yet he could not publicly state his new opinions. The discovery of the "Dear Boy" had brought with it fame and fortune, and if he quickly shifted his ideas again his impetuous nature would again embarrass him. Instead he waited for more evidence, but by 1963 he was itching to announce the new humans to the world.

If the job of describing the fossils sat with Louis alone, he could have published them, but he had brought in Philip Tobias and John Napier to help with the description. Tobias, in particular, thought the bones were from some kind of australopithecine, not, as Leakey thought, an early member of the genus *Homo*, but Leaky eventually convinced him otherwise. Though cultural evidence was not technically acceptable in determining the relationship of fossil humans, Louis, at least, was certain that any human that made tools had to be ancestral to us, and so in 1964 the three published the description of the earliest known member of our genus, *Homo habilis*. The fact that it was found at Olduvai with Zinj had chilling implications. The human that had once been heralded as our proud, tool-making ancestor was now an unfortunate victim of our early relatives:

> While it is possible that *Zinjanthropus* and *Homo habilis* both made stone tools, it is probably that the latter was the more advanced tool maker and that the *Zinjanthropus* skull represents an intruder (or a victim) on a *Homo habilis* living site.

But were humans really such savage competitors in the "struggle for existence"? Although popularized in books like Robert Ardrey's *African Genesis* and the introduction to the film *2001: A Space Odyssey*, Dart's "blood-bespattered" view of our origins was so ludicrously graphic that it failed to gain traction among other paleoanthropologists. Still, it was not far from the consensus that meat eating, tool using, and hunting were what made us human. These were the traits that had allowed our ancestors to rise above the apes. Indeed, the tool making and inferred carnivorous habits of *Homo habilis* fit well within the "Man the Hunter" model of human origins. The acquisition of meat not only provided protein for brain expansion, but it provided selective advantages for hominids to work cooperatively and make increasingly advanced weaponry. As Sherwood Washburn and C. S. Lancaster wrote in a paper presented, appropriately enough, at the "Man the Hunter" conference in 1966, "In a very real sense our intellect, interests, emotions, and basic social life—all are evolutionary products of the success of the hunting adaptation."

Just as with Owen's attempt to separate humans from apes on the basis of the hippocampus minor, however, meat eating, hunting, and tool use would not last long as hallmarks of humanity. What made the difference came not from fossils, but a shift in our understanding of living apes. Prior to the 1960s chimpanzees were seen as peaceful, fruit-eating apes that offered an alternative to our violent ways. The work of

Jane Goodall, who was selected by Louis Leakey to study the chimpanzees at Gombe Stream Preserve in Tanzania starting in 1960, changed all that.[83]

What Goodall observed drastically altered our understanding of chimpanzees. Not only did they hunt and eat meat, but they also made simple tools. Goodall observed chimpanzees carefully selecting twigs and stripping them of leaves in order to "fish" for termites, identifying the apes as not only tool users but tool manufacturers. (Subsequent studies have enlarged the chimpanzee toolkit and even shown cultural differences in the ways tools are used in different populations.) When Goodall telegraphed Louis Leakey with this news, he replied, "Now we must redefine tool, redefine Man, or accept chimpanzees as humans."

While chimpanzees were not regularly having bushpig for dinner or making machines, Goodall's observations threatened the ape-human boundary. Horrific observations made at Gombe in the 1970s would bring them even closer. As in the *Planet of the Apes* mythos, it was widely thought that "ape did not kill ape," at least not unless there was something pathologically wrong with an individual. Conflict sometimes occurred between one group of chimpanzees living in a forest and another, but it was relatively peaceful and did not result in death.

Then, in the early 1970s, a small group of Gombe chimpanzees split from a larger group and took up residence in the jungle nearby. Conflicts between the groups were common, and in 1974 researchers were horrified to see a group of males from one group chase down and savagely beat a lone male from the other. The male died from his injuries, and this was just the first of the fatalities. By 1977, all the males from the smaller group had been killed. Murder, and perhaps even warfare, was not restricted to our species.

While studies of gorillas and orangutans helped to clear misunderstandings about those apes, it was chimpanzees that, in a multitude of ways, threatened notions of human uniqueness. Even so, humans remained taxonomically divided from living apes. Most scientists placed humans within the Hominidae and all apes into the Pongidae, an arrangement similar to Darwin's notebook diagram of 1868. This was reinforced by the notion among paleoanthropologists that the fossil ape *Ramapithecus* was the earliest-known hominid, placing the date of the human-ape split about fifteen to twenty-five million years ago. Comparisons of the differences in the blood-protein chemistry of humans and apes made during the early years of Goodall's studies, however, suggested a much closer family relationship.

In 1967, biochemists Vincent Sarich and Allan Wilson set out to create a "molecular clock" for human evolution. The concept behind it was relatively simple. If there had been steady, regular rates of mutation during the evolution of humans and apes, then it should be possible to look at differences between certain proteins and determine how long it took for those differences to be created. The accumulation of mutations in a lineage of organisms over time would be like the regular ticking of a clock that would allow Sarich and Wilson to count backward to the last ancestor common to apes and humans.

When Sarich and Wilson carried out the experiment they came to a hypothetical divergence date of about five million years ago, and there was no sign of any slowdown in mutations that would have skewed their results. If the biochemists were correct, then the australopithecines sat near the base of our family tree and *Ramapithecus* would be booted out of it entirely. This did not sit well with many paleoanthropologists. For them, the fossil evidence was clear that humans and apes did not diverge earlier than the fourteen-million-year-old hominid *Ramapithecus*. The paleontologists resented that a few lab monkeys were making pronouncements about their field, while many biochemists regarded the paleoanthropologists as fossils who refused to get with the times.

Biochemistry alone could not resolve the issue. Future fossil discoveries were required to confirm or refute the dueling hypotheses. One of the most promising areas for this research was the Rift Valley in northeastern Africa. Pockmarked by active and dormant volcanoes, it was the right place, with accessible deposits of the right age, to potentially yield more fossil hominids. It was in the northern range of this region that the French anthropologist Maurice Taieb discovered a two- to three-million-year-old site at Hadar in the Afar region of Ethiopia in 1970. By 1973, he had organized the International Afar Research Expedition. The American paleoanthropologist Donald Johanson was among the early members of the group.

Early work was difficult, both in terms of fieldwork and camp politics, but toward the end of the first field season Johanson had discovered the upper part of a tibia (or shinbone) and the lower part of a femur. Together they reconstructed a knee of a bipedal creature, and since only humans habitually walked on two legs Johanson was sure he had found a hominid.

When we stand up straight, our femurs angle inward toward the midline of the body. Our hip sockets are set farther apart than where our knees nearly meet beneath our hips. This arrangement is important to balance and allows us to walk upright, and it was these characteristics that had allowed W. E. Le Gros Clark to identify the australopithecines of South Africa as being more humanlike in posture years before. Johanson's find, too, had more in common with us than living apes.

If Johanson was right, his discovery was a major find, but he needed something for comparison and he could not wait to get back to the anatomy lab to check. With Tom Gray, Johanson raided a burial chamber set up by the local Afar people and stole a human femur. This sort of looting ran sharply against the basic ethical standards of science, but despite his theft, Johanson was relieved when the recent bone showed the same orientation as its fossil counterpart.

The discovery of a three-million-year-old bipedal hominid was exciting by itself, but a more momentous find would be made the next year. To the joy of Johanson and his fellow team members, parts of the skull, jaw, arms, legs, fingers, ribs, vertebrae, and hips from a single individual hominid of the same type were found at Hadar. As the scientists reveled that night, the Beatles song "Lucy in the Sky with Diamonds" played endlessly—so it seemed only fitting to name the individual "Lucy."

The skeleton had an informal name, but Johanson was still unsure of just what kind of hominid it was. With the permission of the Ethiopian government he brought the remains back to the United States, where he teamed up with another young paleoanthropologist, Tim White, to pick away at the affinities of the bones. The quandary was not easily solved. It took years to sort out, during which time more fragmentary remains of this kind of hominid came to light. A whole group of the same type of creature—the "First Family"—was found at Hadar. Similar fossils were also found at Laetoli in Tanzania, where White had worked with Mary Leakey.

The large collection of bones showed a mosaic of traits, similar to both later humans and to apes, and the specimens did not fit neatly into the known varieties of *Homo* or *Australopithecus*. This caused nearly ceaseless arguments between Johanson and White, and among both of them and Mary Leakey, but by the fall of 1977 the duo knew they had found a new species. They decided to call it *Australopithecus afarensis*. It was a hominid with legs and hips suited to walking bipedally but an upper body retaining adaptations to moving through the trees. It had long arms, curved fingers, and a funnel-shaped ribcage like living apes,

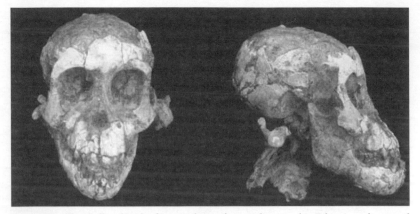

FIGURE 91 – The skull and neck of a juvenile *A. afarensis* discovered in Ethiopia and reported in 2007.

but it also had hips and legs that indicated an upright posture. This was reinforced by the presence of the track way of a bipedal hominid at Laetoli, where they found type material for *A. afarensis*, which seemed to be a likely candidate for the track maker. Clearly, this was a creature set to shake up the evolutionary tree.

Ever since the scientific recognition of the Neanderthals, experts had debated whether one type of hominid or another was a true ancestor of our species or represented a side branch that left no living descendants. Personal philosophies influenced these arrangements as much as the bones did, from the early rejection of australopithecines as hominids to Louis Leakey's dogged search for the earliest members of the genus *Homo*. *Australopithecus afarensis*, though, appeared to occupy a critical junction in human evolution. It was older than the other australopithecines (*A. africanus* and *Paranthropus*) and *Homo habilis*, yet it showed resemblances to both lines. *A. afarensis* seemed to occupy a place near the split of later australopithecines and the earliest members of our own genus. Given that *A. afarensis* was only about three million years old and exhibited so many apelike traits, though, it hinted that our forebears diverged from the ancestors of living apes more recently than many paleoanthropologists were prepared to accept.

The transitional status of Lucy was becoming established just as *Ramapithecus* was being torn down from its vaunted position. When the first jaw fragments of the supposed hominid were found they were thought to take a distinctively human shape. In apes, the molars and premolars form a straight line with the incisors and canines slightly curving to give the jaw a rigid U shape. In humans, however, the rows and molars and premolars are more curved and have a parabolic shape. The fragments of *Ramapithecus*'s jaw seemed to fit the human pattern, but when more complete fossils of the ape were found in the 1970s it turned out that the fossils called "Ramapithecus" belonged to another creature.

In many primate species males and females are highly sexually dimorphic, meaning that they differ significantly in traits like size, skeletal structure, canine tooth length, musculature, and other features. Orangutans and gorillas are good examples of this; if you looked at the skulls of a male and female gorilla side by side and did not know any better you could be excused for thinking they belonged to two distinct species. Rather than being products of natural selection, these differences are the result of sexual selection—such as female choice and the competition between males for mates—and this would have held true for ancient primates, too. When more complete skulls of *Ramapithecus* came to light it became apparent that they represented the more gracile (and probably female) form of *Sivapithecus*, a fossil ape that closely resembles living orangutans. This revision made it no longer reasonable to push the divergence of humans and apes back past fourteen million years.

The fall of *Ramapithecus* only represented half of the major shift occurring in anthropology, however. At about the same time, molecular biologists Charles Sibley and Jon Ahlquist were pioneering a new technique called DNA hybridization to determine the relatedness of species. In this technique, segments of DNA from two species are hybridized with each other, and the more similar the DNA strands of the two species are the more closely they will match (indicated by how much energy it takes to split them apart again). The technique had worked well enough on birds that Sibley and Ahlquist decided to try it on primates. Not only did they pin the human-ape divergence as occurring seven to nine million years ago, but they confirmed that chimpanzees were more closely related to humans than they were to other apes.

Further studies showed that the difference between the genetic codes of humans and chimpanzees was very small, only a few percent, but scientists were even more surprised when they found our "missing" chromosome. Chimpanzees, gorillas, and orangutans have twenty-four pairs of chromosomes and we have twenty-three. Since we shared a common ancestor with these apes it is likely that our ancestors also had twenty-four pairs. At some point, the chromosome number in the human lineage was reduced to twenty-three, but rather than disappearing entirely the "missing" chromosome still remains in our cells. It is part of our chromosome 2. At some time during our evolution the "missing" chromosome fused to the end of chromosome 2, and so the end of the original chromosome 2 became the middle of the new, fused one. Not only does our present chromosome 2 show that such an event occurred, but the entire chromosome is strikingly similar to the chimpanzee chromosomes 2p and 2q combined.

In light of all this new evidence the classic division between the Pongidae and Hominidae could no longer be upheld. Our species was situated *within* the ape group, not outside of it, and our place in nature had to be reconciled with the new evidence.

Our species, and all of our extinct relatives more closely related to us than to chimpanzees, belong to the hominin lineage. Since chimpanzees, belonging to the panin line, are our closest living relatives, the two lineages together constitute the group Hominini. This group is itself nested in a larger group that includes the gorilla and orangutan lineages called the Hominidae, and the Hominidae is part of an even more inclusive group, of which gibbons are a part, called the Hominoidea. These terms are confusingly similar, but each rank represents differing degrees of specificity that relate all apes to each other. We are hominins, but, as far as our family tree is concerned, we are also hominids and hominoids.

By the 1990s much had changed in paleoanthropology. Our species was identified as being a part of the ape family tree, not outside it, and the preoccupation with "Man the Hunter" had faded. Even Dart's horrifying view of the past had been permanently laid to rest. In his investigation of the Sterkfontein and Mapakansgat caves, the paleoanthropologist C. K. Brain discovered that the bone, tooth, and horn "tools" Dart found had actually been made naturally. Most of the remains in the caves had been washed in or deposited by carnivores, causing wear that Dart mistook as signs of tool use. The weight of the evidence revealed that the australopithecines had probably been prey more often than predators, particularly the skullcap of a young *Paranthropus* that had two puncture marks in it indicating that it had been killed by a leopard. It was only in the younger cave deposits, from when the caves were inhabited by *Homo erectus*, that clear signs of human hunting (like herbivore leg bones bearing tool cut marks) were found.[84]

Homo erectus, like its australopithecine relatives, also received a makeover in the last quarter of the twentieth century. When the first remains of *Homo erectus* were found by Dubois ("Pithecanthropus") they were thought to have belonged to a creature intermediate between human and ape. As the discoveries from Dragon Bone Hill ("Sinanthropus") and sites from Africa were compared, though, it became clear that *Homo erectus* was more similar to us than our common ancestor with chimpanzees. Key to this revision was a specimen found by Kamoya Kimeu near Lake Turkana, Kenya, in 1984.

Kimeu was part of a team working with Richard Leakey, son of Louis and Mary, at a 1.5-million-year-old site called Nariokatome when he discovered the nearly complete skeleton of a male juvenile *Homo erectus*. Nicknamed "Turkana Boy," this nine to ten year old was already five feet, six inches tall, putting him head and shoulders above his australopithecine ancestors. He also had a relatively narrow pelvis and lanky proportions, quite different than the stooped images of "Java Man" from early twentieth-century books and magazines.

Even though the Nariokatome skeleton was of a juvenile male, its hips provided paleoanthropologists with a starting point for determining the pelvis shape of adult female *Homo erectus*. This was important, as in order to understand how our species was born we had to know how individuals were born at different times in our ancient history. As our lineage evolved larger brains, the increasing size of the infant head had to be accommodated by the hips, and in our species birth is extremely painful and stressful due to a convoluted birth canal and the large size of the infant skull. Even though the infant skull is not fully fused and can distort somewhat, it is still a tight squeeze.

Apes, however, do not go through the same discomfort human mothers do. They have a straight birth canal and give birth with relative ease. The differences continue after birth, as well. Since we have such large heads we have to be born "prematurely" to make sure the skull will fit through the birth canal; we are helpless for a much longer time than infant apes are. If we developed any longer in the womb, birth might not be possible, though, and so a long period of helplessness after birth is a trade-off for larger adult brains. The narrow-hipped *Homo erectus* appeared to have a pattern of birth and growth similar to our own, with a longer period of infant helplessness outside the womb.

What was needed to test this idea, however, was the pelvis from an adult female *Homo erectus*. It remained elusive for years, but in the fall of 2008 Scott Simpson and colleagues announced that they had found the coveted prize from 1.2-million-year-old sediment of Ethiopia. Given that *Homo erectus* was closely related to our species, it was possible to use the same anatomical landmarks in the hips that identify human females, so the researchers could be sure they had sexed the bones properly.

The newly discovered *Homo erectus* hips were much wider than those of the Nariokatome skeleton, and even of modern *Homo sapiens* females. These hips housed a more capacious birth canal and would have allowed for an infant head 30 percent larger than previously calculated to pass through, making it probable that *Homo erectus* children were born relatively more developed than *Homo sapiens* infants and spent a shorter amount of time being helpless. The pelvis also showed that there were probably some stark differences between *Homo erectus* males and females. The males may have been tall and lanky like Turkana Boy, while the females might have been shorter and wider. (More specimens will be needed to test this and make sure it is not simply a case of regional variation.) This sexual dimorphism and a growth rate intermediate between that of our species and living apes emphasizes our gradual, mosaic evolution from ancient apes.

Linnaeus was right. There is no character or trait that can be zealously wielded to obliterate our blood relationship with other primates. We are apes, just of a different sort, and many of our traits can be traced back millions of years into the past. Opposable thumbs, fingernails, binocular vision, and many other traits that we think of as distinctively human evolved among primates during the sixty-five million years after the extinction of the dinosaurs, and the first members of our own hominin lineage between five and seven million years ago were little different from the apes from which they evolved.

In 2001, Brigitte Senut and Martin Pickford announced that the hominin *Orrorin tugenensis* had been discovered in the Tugen Hills of Kenya. The leftovers from a carnivore kill (as evidenced by toothmarks on the bones), there was not much left of *Orrorin*, but enough of its skeleton was recovered to tell that it was relevant to early human evolution.[85] At 5.8 to 6.1 million years old, *Orrorin* falls into the proposed time range during which the last common ancestor of chimpanzees and humans would have existed, and it appears to have possessed a key human trait.

Among the *Orrorin* bones was the top portion of a femur, including the surface that articulated with the hip. This bone was very similar to the femurs of australopithecines in having a long femoral neck that would have dictated where the gluteal muscles from the hip attached to a knob of bone called the greater trochanter. This is important, as the gluteal muscles help us keep balanced as we walk, and from the available evidence *Orrorin* probably stood upright and walked in a manner similar to that of the australopithecines.

Another, even older, potential early human was described the following year. During the latter part of the twentieth century it was

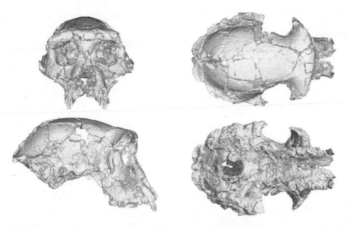

FIGURE 92 – The virtual reconstruction of the skull of *Sahelanthropus* to correct for the distortion seen in the original skull (clockwise from upper left: the skull as seen from the front, top, right side, and bottom). Note the placement of the foramen magnum directly beneath the skull in the view of the skull's underside.

thought that Africa's Rift Valley was the crucible of human evolution, as all the earliest hominins had been found there. But in 2001, early hominin fossils were found far to the west in the Sahel desert of Chad. Discovered by Djimdoumalbaye Ahounta, a local man working with several of his countrymen and French paleoanthropologist Alain Beauvilain, the creature was represented by a crushed, but complete, skull, part of a jaw, and a few teeth. It was dubbed *Sahelanthropus tchadensis*, and it caused quite a stir in the media and scientific circles. Its crushed skull soon appeared in magazines, on book covers, and on television programs as a definite human ancestor. This fascination with *Sahelanthropus* is not that surprising. Whereas the other apes were represented by bits and pieces, *Sahelanthropus* had a face, and it was said to be seven million years old. This would make it the oldest hominin yet known, but the interpretation of *Sahelanthropus* is not so straightforward.

The skull of *Sahelanthropus* was not found embedded in sediment, but on the wind-blown surface of the desert. This makes its age questionable, as the winds are so powerful they can transport and even destroy fossils. Its identification as a hominin is also tenuous. Even though much was initially made of its flat face, a reconstruction undertaken by Cristoph Zollikofer and colleagues in 2005 revealed that *Sahelanthropus* did have a lower face that jutted out. (The announcement of a twelve-million-year-old, flat-faced fossil ape *Anoiapithecus* from Spain in 2009 showed that a flat face was not a useful trait in distinguishing early humans in any case.) Although the reconstructed skull was not unlike those of other hominins, the skull of *Sahelanthropus* also closely resembled that of a fourteen-million-year-old ape from Spain named *Pierolapithecus*. Both had straight lines of molars and premolars in the jaw, like non-human apes, and the faces of the two genera are very similar. Given that *Pierolapithecus* has been suggested as being close to the common ancestor of gorillas and hominins, it is possible that *Sahelanthropus* represents a non-human ape that predates the chimpanzee-human split.

But there was yet another early hominin, named but not fully described, that was of high importance to the debates over the earliest humans. During the early 1990s a team led by Tim White discovered the fragments of a new kind of hominin in the 4.4-million-year-old strata of the Afar Depression in Ethiopia. It was briefly described as *Ardipithecus ramidus* in 1994, but everyone knew that White's team had found much more. The trouble was that most of the fossils, including a crushed skull, were extremely delicate and required a lot of preparation, and so (somewhat like Dubois's *Pithecanthropus*) they were studied in secret for a decade and a half.

The importance of *Ardipithecus ramidus* was played up even before its full description. With the description of a related species, *Ardipithecus kadabba* in 2004, the Afar fossils were groomed to fall in a straight line of early human evolution. It would have started 5.6 million years ago with *Ardipithecus kadabba*, giving way to *Ardipithecus ramidus* by 4.4 million years ago, leading into another new hominin called *Australopithecus anamensis* by 4.2 million years ago, which transitioned into *Australopithecus afarensis* around four million years ago. Some paleoanthropologists took this temporal arrangement and the proximity in which these species were found to suggest that there was a single, linear march of these early hominin species in northeastern Africa, with a major split only occurring after the evolution of Lucy and her kind.

This was the context through which *Ardipithecus ramidus*, nicknamed "Ardi," was introduced to the public on October 2, 2009. An entire issue of the journal *Science* was devoted to the fossils, covering everything from what sort of animals lived alongside Ardi to inferred social behaviors of our 4.4-million-year-old relative.[86] But what was most fascinating was that *Ardipithecus ramidus* did not look like something approaching a chimpanzee. It was a different kind of ape that ran counter to popular perceptions of human evolution.

The origin of bipedalism has been a vexing question for anthropologists for decades. No other mammal walks like we do, and even our closest living relatives (gorillas and chimpanzees) are slouched over to walk on their knuckles. According to different hypotheses, early humans stood up to see predators, to carry tools, to pick fruit, to hold babies while walking, to be better long-distance runners, among other activities—with the focus always being on some evolutionary pressure that made us stand up. Since chimpanzees were taken as a general model for what our ancestors might be like, it was assumed that the progenitors of the first humans would have been knuckle-walking apes, too; there had to be some powerful explanation for the shift.

The skeleton of *Ardipithecus ramidus* suggests something very different. It had a combination of traits such as long arms, curved fingers, and a divergent big toe that point toward a life in the trees. Furthermore, based upon the recovered pelvis fragments, the hips of *Ardipithecus ramidus* were probably wider and more bowl-shaped than what is seen in chimpanzees. And while the ends of femur that would make up the knee joint was missing, the upper leg bone did have an inward slant suggestive of the arrangement seen in fully bipedal hominins such as *Australopithecus afarensis*. Further down, a tiny bone of the foot called the os perineum also provided an interesting contrast with living apes. In monkeys and other primates this tiny bone is embedded in a tendon that allows them to keep the foot rigid, which is important to securely landing on a branch when jumping from one tree to another. Chimpanzees and gorillas lack this "rigid foot" feature, however, and the fact that *Ardipithecus ramidus* had it might have made an important difference in the way it moved.

Altogether the anatomy of *Ardipithecus ramidus* is suggestive of an ape that lived in the trees and got around by grasping branches with both its hands and feet. It did not show adaptations in the wrist or hips to being a knuckle walker as seen in living chimpanzees and gorillas, but the adaptations *Ardipithecus ramidus* had for life in the trees might have allowed it to also walk on two legs when on the ground. Keeping a rigid foot would have helped the hominin keep its balance, just as the shape of its femur and hips would. These small differences might have made it easier to stand and move upright while on the ground than to move around on all fours,so bipedalism (rather than knuckle-walking) evolved among hominins.[87] This means that our upright posture would be an exaptation: an adaptation that was jury-rigged from already existing anatomical traits for a new purpose.

But *Ardipithecus ramidus* is not the only early hominin to show hints of bipedalism (or traits that could be co-opted to compel upright walking). *Orrorin* possessed modifications of the femur closely tied to walking upright, and the position of the foramen magnum undernearth the skull of *Sahelanthropus* also hints that when the rest of its skeleton is found it may express other traits indicative of bipedalism. While the fossils currently known did not overlap in time they could all represent an early diversification of arboreal hominids that possessed traits that could be easily co-opted for life on the ground, which also makes it possible that there might have been bipedal apes not directly ancestral to us. Take *Sahelanthropus*, for example. Let's assume that the rest of its skeleton will show other traits associated with bipedalism but that it turns out to be a hominid, not a hominin. This would mean that, when looking at fossils from the time the first humans were thought to have evolved (still between about seven and five million years ago), we could no longer assume that an ape with bipedal traits was automatically an early human. As with other groups of fossil organisms careful comparisons would be required to tease out the relationships of early fossils based upon shared, derived characteristics, and it is entirely possible that early members of any group did not express the definitive characteristics seen in later members of the group. To put it more simply, there would be no single trait that would draw a stark line between "ape" and "human," and focusing too heavily on bipedalism might obscure matters more than illuminate them.

The systematic status of *Ardipithecus*, *Orrorin*, and *Sahelanthropus* will no doubt remain controversial for some time, but these new genera illustrate the complex nature of human evolutionary history. Hominins, like the anthropoids and apes, appear to have undergone an early diversification, and those so far discovered hint that there are certainly other as-yet-unknown early hominin genera waiting to be found. Even though the news media, and some anthropologists, seem obsessed with finding our direct ancestors, the reality presented by the fossil record is much more complicated, and it is highly unlikely that we have yet discovered our earliest hominin ancestor. The evolutionary history of hominins is best viewed as a branching bush, not a straight line of ascent.

The general pattern of hominin evolution as presently understood looks something like this. What *Ardipithecus*, *Orrorin,* and *Sahelanthropus* may indicate is that there was a radiation of hominin types around six million years ago. That diversity appears to have given way to a bottleneck by about 4.2 million years ago, at which time *Australopithecus anamensis* was living in the woodlands of East Africa. One population of these hominins may have given rise to *Australopithecus afarensis*, which is the only known hominin present

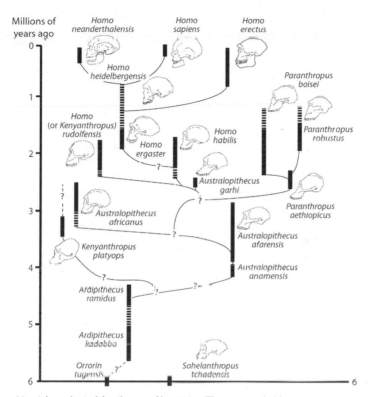

FIGURE 93 – A hypothetical family tree of hominins. There are probably many more species yet to be discovered, and in the coming decades some of these relationships may change as others become better supported. Even so, the overwhelming picture is that humans have had a "bushy" evolutionary history that has left only one living species.

from about four to three million years ago. After that, hominin evolution branched out again.

On the one side there was *Australopithecus africanus*, a gracile australopithecine related to the robust forms like *Paranthropus aethiopicus* and *Paranthropus robustus* (akin to the Leakeys' "Dear Boy"). These robust australopithecines resembled their gracile predecessors in overall form, but they had massive teeth and jaws that might have allowed them to consume a wide variety of food in the open, grasslandlike habitat in which they lived. This would have been especially important during harsh seasonal changes when there may not have been much other than tubers to eat. The first of their kind appeared 2.6 million years ago and persisted until about one million years ago, but ultimately they left no descendants.

The other side of the split, which included our ancestors and close relatives, was also undergoing a diversification around 2.5 million years ago. At that time, the hominin *Australopithecus garhi* lived in what is now Ethiopia, temporally overlapping with *Australopithecus africanus* and possibly the earliest members of our genus, *Homo*. It was not one of our ancestors, but a close relative, who quickly disappeared as hominins like *Homo habilis* came on the scene. At places like Olduvai Gorge, in fact, *Homo habilis* lived alongside other hominins like *Paranthropus boisei*, and there were indeed several kinds of human running around Africa at once. What had caused this proliferation may have been a change in climate. From about 2.8 to 2.4 million years ago there was a climatic pulse that caused the further spread of open grassland habitats dominated by a cooler, drier climate. This change is correlated with the split in the human family tree, and the ecological shift may have created new selective pressures that led to the flowering of hominin species.

This branching pattern is exactly what would be expected if natural selection was a primary mechanism in evolution. Entire species do not follow a "March of Progress" from a "lower" form into a "higher" one; evolution happens due to differences in gene flow and selective pressures on different populations. In this way, one population can remain relatively unchanged while another undergoes rapid evolution, producing a new species that overlaps in time with their ancestral species. This applied to hominins as to any other organism.

In 2007, Fred Spoor and his colleagues reported that the 1.4-million-year-old site of Ileret, Kenya, presented evidence that *Homo habilis* and *Homo erectus* had coexisted there for about 500,000 years. Media outlets were thrown into a tizzy over this discovery, and the find was mistakenly said to show that *Homo erectus* could not have evolved from *Homo habilis*. How could ancestors live alongside their descendents?

The truth of the matter is that the paper confirmed that evolution is a branching process that occurs in populations that may overlap with each other temporally and even spatially. If the two species really do constitute an ancestor-descendent relationship then a population of *Homo habilis* gave rise to *Homo erectus*, perhaps as a result of another climatic pulse that led to the further expansion of grasslands around two million years ago. But other populations of *Homo habilis* persisted for a time before dying out around 1.6 million years ago. This fits the punctuated equilibrium pattern of evolution predicted by Eldredge and Gould, wherein there is a rapid evolutionary change, resulting in a new species, followed by little to no change at all.

By the time *Homo habilis* perished *Homo erectus* was already a world traveler. The *Homo erectus* bones from Java are as ancient as the oldest *Homo erectus* from Africa, and the discovery of 1.9-million-year-old *Homo erectus* skulls from Georgia have shown that this hominin dispersed from Africa early.[88] There may have been more than one wave of dispersal, but *Homo erectus* quickly spread through Africa, the Middle East, Asia, and Indonesia. (Much like the Channel Islands mammoths, one population of *Homo erectus* in Indonesia appears to have become dwarfed and adapted into a different species, *Homo floresiensis*.) It was a species adapted for life in open landscapes, and due to their increased energy needs (which included a larger brain) *Homo erectus* required higher quality foods. It was among the first humans to regularly obtain meat through hunting, and later individuals of the species may have been among the first humans to use fire for cooking.

Eventually, however, the widespread populations of *Homo erectus* died out, with the last population in Asia persisting until about 200,000 years ago. Before that occurred, though, a population of *Homo erectus* gave rise to a species called *Homo heidelbergensis* that appeared in Europe about 600,000 years ago. This is the probable ancestor to our close relative *Homo neanderthalensis*, the Neanderthal, which evolved in Europe about 130,000 years ago.[89] They were robustly built with deep, chinless lower jaws and low-domed skulls that held brains as large, if not larger, for their body size than ours. They made advanced tools, buried their dead, and perhaps even made music; Neanderthals were much more intelligent than the club-wielding imbeciles they are often portrayed as. By 30,000 years ago, however, they had disappeared.

No one is sure what happened to the Neanderthals, but our own species has always been a prime suspect for their extinction. During the time that Neanderthals evolved in Europe, our species was evolving in Africa, sometime between 200,000 and 100,000 years ago. At 70,000 years ago, one population left Africa and, like *Homo erectus*, our species spread over the globe. There was not a separate origin for different races of *Homo sapiens* in different parts of the world, but a common African ancestry that links all living people together.

Ancient members of our species in Europe, the Cro Magnons, began their habitation there about 40,000 years ago and may have come into contact with the Neanderthals. Did we decimate their populations in combat? Did we carry diseases that wiped them out? Did we hybridize with them (making us part Neanderthal)? Or was it something else entirely, like climate change, that did in our sister species? No Neanderthal skull has been found with a Cro Magnon spear point in it, so it is difficult to tell how *Homo sapiens* and Neanderthals may have competed. Neither have supposed hybrid skeletons held up to scrutiny, and while a study published in early 2010 hints that we mayl have interbred with Neanderthals, this hypothesis requires further study and does not explain their extinction. Given that the Neanderthals survived the harsh climatic changes of the Ice Age, it also seems unlikely that they would have died when the glaciers began to recede. The reason for their extinction remains elusive, but when they did disappear it was the first time in millions of years that there was only one kind of human.

In our lonely isolation we have often wondered, "What makes us human?" We now know the answer: "Surprisingly little." The remains of extinct hominins have exquisitely documented our family history, and our understanding of living apes has further assaulted every fortification meant to defend our uniqueness. From the conformation of our bones to the intricacies of our genetic code, our bodies testify to our ancient history.

I can only wonder how we would perceive ourselves if the Neanderthals or robust australopithecines survived today. Would they help us better understand our origins, or would we destroy them out of revulsion? Perhaps it is best that we will never know. We may want to identify the epitome of "humanness," to find comfort in some unassailable characteristic that makes us superior (be it in the eyes of God or our own), but there is no bright line dividing us from the apes. Since we are apes ourselves, I do not expect one to be found. For to ask "What makes us human?" assumes that there was sa single instant, hidden in the past, in which we transcended some boundary and left the ape part of ourselves behind. We forget that those are labels we have created to help organize and understand nature, as if sometime in the Pleistocene there was a glorious moment when language, art, and culture swept into the world and gave us dominion over it.

Even scientists have sometimes searched for an evolutionary hero and constructed tales of how we came "up from the ape." The honest desire to find such a character will ultimately be disappointed. There was never an "ascent of man," no matter how desperately we might wish for there to be, just as there has not been a "descent of man" into degeneracy from a noble ancestor. We are merely a shivering twig that is the last vestige of a richer family tree. Foolishly, we have taken our isolation to mean that we are the true victors in life's relentless race.

Whether meaning is to be found in the heavens or in ourselves, we feel a pervasive need to ennoble our heritage. What else have we if we do not? History tells us that we are the descendants neither of an ape that intentionally strove to reach higher cerebral branches nor a holy couple created by divine fiat. Instead we are inheritors of a rare intelligence that can permeate the delicate workings of nature but fears what it might find there. There is no reason to fear. Life is most precious when its unity and rarity are recognized, and we are among the rarest of things.

Time and Chance

I returned, and saw under the sun, that the race is not to the swift, nor the
battle to the strong, neither yet bread to the wise, nor yet riches to men of
understanding, nor yet favour to men of skill; but time and chance hap-
peneth to them all.   —Ecclesiastes 9:11 (King James Version)

Many years ago, when I was about five, my parents took me to see the
dinosaurs at the American Museum of Natural History. This was before
the great renovation of the paleontology exhibits in the 1990s. When I
visited, the museum's iconic *Tyrannosaurus* reared back as if doing a
Godzilla impression, and the "Brontosaurus" loomed over the Jurassic
Dinosaur Hall, complete with the wrong head. I was enthralled by the
enormous skeletons that stood in the dim fossil halls, and as I stood in
their long shadows my imagination ran away with me. I wondered what
those old bones would have looked like when clothed in flesh, and in
one fleeting moment, left alone in the quiet of the hall with my family,
I could have sworn that I could just make out the dinosaurs' distant
rumbles and growls echoing across the depth of time.

I still visit the museum's dinosaurs every few months or so. Some of
the childhood whimsy is gone, but they continue to fascinate me.
Beyond the awesome grandeur of the reconstructed skeletons of *Gorgosaurus*, *Triceratops*, and other long-lost creatures, every single bone
has a story to tell about the life and evolution of the animal it once
belonged to. The skeletons of the dinosaurs are much more than static
monuments of a bygone age; they are intricately detailed records of an
ongoing evolutionary "experiment" that has been carried out on this
planet for the past 3.5 billion years and provide the context for understanding our own evolutionary history.

Perhaps, so long after the birth of paleontology, it is easy to forget
what we have learned from such creatures, but by studying their
remains we have begun to better understand the world we live in, from

the fact of extinction to how life on earth came to be as it is today. Indeed, the history of the dinosaurs is particularly instructive because the mass extinction that wiped them out, among other creatures, sixty-five million years ago opened up new evolutionary opportunities for mammals, including our own early primate ancestors. But what if the asteroid that triggered the mass dying had missed the earth entirely? What would life look like today if, by some quirk of history, the dinosaurs had survived?

An extended tenure for the dinosaurs might appear to be a subject better suited to science fiction than academic debate, but in 1982 pale-ontologist Dale Russell presented the "Dinosauroid,"—a scientifically derived vision of what a modern-day descendant of the maniraptoran dinosaur *Troodon* might look like.[90] With its bipedal stance, grasping hands, forward-oriented eyes, and relatively large brain, *Troodon* looked to Russell to be evolving towards a humanlike condition. Since Russell believed that our species possesses the only body capable of housing our complex minds, he affirmed that the lineage of *Troodon*—had it con-tinued to evolve high levels of intelligence—would have been molded into something like us. With the help of artist Ron Seguin, Russell cre-ated a full-sized sculpture of his thought experiment, the end result being a creature that looked a more refined version of the sleestaks that men-aced the heroes of the 1970s television series *Land of the Lost*.

Though it would be a stretch to call the Dinosauroid beautiful, it did stroke our vanity. It was an assurance that evolution guaranteed the arrival of something like us, if not exactly our species, by way of natural law. Never mind that we now know that close relatives of *Troodon* survived the mass extinction in the form of birds (some of which, such as crows, are highly intelligent animals with an anatomy very different from our own). The Dinosauroid was an attempt not so much to ask what life might be like now had the extinction been can-celed but to affirm that there is a sort of "human niche," which would inevitably be filled by one kind of creature or another in time. We could feel confident that our species had practically been foreordained from the time the first life emerged. Given how easily this can be squared with the beliefs of human origins as told by Abrahamic religions it is not surprising that a number of scientists who hold theistic beliefs (such as Russell, Francis Collins, Kenneth Miller, and Simon Conway Morris) have advocated similar versions of human inevitability. Rather than being open-ended, evolution would be a process so constrained by the laws of physics and chemistry that, if we had the ability to restart evo-lution from the beginning, the history of life would end up looking much the same. In this light it could be believed that the emergence of *Homo sapiens*, the single remaining species of hominin on a backwater planet far removed from the center of the galaxy or even the universe, was woven into the very fabric of the universe itself.

But our human conceit blinds us to the true pattern of the fossil record. The diversity of life and the disparity between organisms cannot solely be explained by a lawlike evolutionary mechanism driving organisms up a ladder of predictable progress. Instead, the fossil record has revealed the growth of the tree of life to be much more haphazard; a directionless unfolding greatly influenced by phenomena from catastrophic mass extinctions to the emergence of new forms thanks to the co-option of the old structures into new ones. Tetrapods carry that name because of an ancient duplication error which caused some fish to have four fins instead of two; the inner ear bones of mammals evolved because of the close connection between ear and jaw in their ancestors; whales swim in their peculiar fashion because of the way their ancestors moved on land; and many of the features which allowed birds to take to the air evolved first in ground-dwelling dinosaurs. These changes were not effected in a linear fashion, as if creatures were striving for a particular goal. Instead, as shown by the evolutionary radiations of fossil elephants, horses, and humans, species branch out over time, oftentimes with descendants living alongside their ancestors. As time goes on, some of these branches are pruned back by extinction while others persist. In our case, especially, the fossil record has unequivocally shown that what was once a diverse family of humans has been reduced to just one species, leaving us good reason to wonder what might happen to our kind in the future.

This view of life, in which we owe our existence to a long series of antecedent states molded by unpredictable events, has not often been welcomed. It cuts to the core of what we have traditionally held up as human exceptionalism by inextricably tying us to quirks of history which could have gone another way. Had the first land-dwelling vertebrates six legs instead of four, had the Permian extinction extirpated the cynodonts, had the asteroid which struck earth sixty-five million years ago missed, or had our ape ancestors clung to an arboreal lifestyle in equatorial forests—there can be no doubt that life on earth today would be very different. If we could somehow rewind and replay earth's history, as suggested by Stephen Jay Gould in his book *Wonderful Life*, we would not expect life to follow the same history with which we are now familiar. Entirely different organisms would probably evolve and perish over the course of history, and the appearance of anything like our species evolving would be extremely slim. Unfortunately we cannot actually run Gould's thought experiment, nor have we yet discovered life elsewhere in the universe with which to compare organisms from our own planet, but an ingenious laboratory experiment has suggested that history and happenstance have key roles to play in evolution.

In 1988, the biologist Richard Lenski founded twelve populations of identical *Escherichia coli* bacteria in his lab at the University of Michigan. To keep track of how the populations were changing, the scientists in the lab froze representative samples of the populations every 500 generations (or about every two and a half months). This provided them with a preserved evolutionary record of each group that could be revitalized to see if evolution would repeat itself from a given starting point. By 2008, over 44,000 generations of the bacteria had lived and died in the lab, and it was in that year that Lenski (along with co-authors Zachary Blount and Christina Borland) reported on a major change that had recently occurred in the populations.

From the beginning of the experiment, the *E. coli* populations were kept in environments with only the minimum materials needed to keep them alive, which also included a resource called citrate. Although citrate can be an important component of the citric acid cycle in cells, *E. coli* cannot transport it across their cell membranes when oxygen is present, so while each of the twelve populations was kept with a store of this useful resource none of them could actually make use of it. As had been expected, the starting populations of *E. coli* were not able to utilize the citrate in their environment, and even though the bacteria were accumulating mutations none gave any of the populations the variability needed to use citrate. By the 31,500 generation mark, however, one population had evolved the ability to use the citrate—but what had caused this change? Did this happen because of a single-point mutation that immediately opened up new possibilities, or did the mutation build upon previous changes to allow the bacteria a new ability?

To find out, the researchers went back to their library of bacteria. Using clones from twelve different points in the study, for a total of 72 "replay" populations in all, the scientists let the bacteria reproduce for about 3,700 generations, checking along the way to see if citrate-using populations evolved again. What they found was that the origin of citrate-using bacteria was contingent upon a series of chance events. The first was a mutation that modified the "genetic background" of the bacteria so that a later mutation would be more likely to produce a citrate-using form. It had gone undetected as it did not have an immediately visible effect on the bacteria. A second mutation, around generation

30,000, allowed some bacteria to start using citrate, and soon afterwards another mutation made the citrate-using bacteria more efficient at using the resource (while simultaneously reducing their ability to utilize glucose as their ancestors in the first generation did).

Even though the ability to use citrate would have been advantageous for any population of bacteria, it took well over a decade, more than 30,000 generations, for that adaptation to appear, and thus far it has only appeared in a single population out of twelve. It was not caused by an extremely rare single event but was precipitated from a succession of prior conditions that allowed the change to take place. Since these changes did not appear in the other eleven populations of bacteria, all grown under identical conditions from the same starting point, the discovery suggests that evolution does not run according to the same path every time. The element of chance, as embodied by mutations and other genetic changes, can open or close potentials for organisms that are then molded by natural selection. As the authors of the study concluded:

> Our study shows that historical contingency can have a profound and lasting impact under the simplest, and thus most stringent, conditions in which initially identical populations evolve in identical environments. Even from so simple a beginning, small happenstances of history may lead populations along different evolutionary paths. A potentiated cell took the one less traveled by, and that has made all the difference.

Unfortunately, we do not have a frozen store of early hominins, Permian synapsids, Devonian tetrapods, or other prehistoric vertebrates ready to be thawed out to run similar experiments, but what was found in the Lenski lab certainly applies to the fossil record. As the biologist Jacques Monod is reported to have said, "What is true for *E. coli* is also true for the elephant." (And also for us!) Small, unpredictable changes can have major influences on how life evolves, cordoning off some evolutionary potentials and allowing for others. Organisms are constrained not only by the chemistry and physics on this planet, from gravity to the composition of the atmosphere, but by the role of chance in their own history.

The complementary roles of contingency and constraint in life's history are among the central lessons of the fossil record. Looking back from our narrowed present perspective we can make out a branching pattern, marked by fits and starts, in which each successively older slice of life's history only makes sense in the context of what came before it.

It is comprehensible, but not arranged according to a predetermined plan, and there is no sign that evolution is being pushed along blessed roads of increasingly perfect adaptations put in place from the very beginning. Life as it is now did not have to be.

We are constantly grappling with our place in nature. In the over 4.6-billion-year history of our planet and the over three billion years that life has existed on earth, organisms like us and our hominin cousins have only evolved once, and at a relatively late date, at that. Rather than suggest that creatures like us were somehow predestined to be, this fact of the record means that, much like the bacteria in Lenski's lab, hominins are the rare products of a contingent process that probably would not evolve again if evolution were restarted. Some might find this hard to accept. Life-forms in innumerable splendid varieties have flourished and been extinguished on this planet, and yet we still feel compelled to find some reason, any reason, that our existence was premeditated and imbued with special meaning. No one argues the case for the inevitability of maple trees or damselflies or aardvarks; it is always our species that is cast as the specially intended pinnacle and purpose of nature. Such hubris is absurd. As Mark Twain jested in his essay "Was the World Made for Man?":

> Man has been here 32,000 years. That it took a hundred million years to prepare the world for him is proof that that is what it was done for. I suppose it is. I dunno. If the Eiffel Tower were now representing the world's age, the skin of paint on the pinnacle-knob at its summit would represent man's share of that age; and anybody would perceive that that skin was what the tower was built for. I reckon they would, I dunno.

If we can let go of our conceit, we may find that an understanding of evolution makes life on this planet all the more precious. Every living species is a unique, still-changing part of lineages that have persisted for billions of years, and once they are gone they are lost forever. The same is true of our species. Sooner or later, extinction claims all.

Nothing quite like us has ever existed on earth before and may not ever again after we are gone. Given the contingencies of our own history, that we exist at all is amazing. If we wish to know ourselves, we must understand our history. We are creatures of time and chance.

Notes

[1] Thankfully the destruction of the Messel pit was stopped. The quarry was given the status of a UNESCO World Heritage site in 1995. Private collectors are no longer allowed to remove fossils, but scientific research continues.

[2] A few journalists, such as *Science* anthropology writer Ann Gibbons, were able to wrangle a look at early versions of the paper, but they had to sign non-disclosure agreements before doing so.

[3] This conclusion was supported by a second paper—submitted for publication before publication of the description of *Afradapis*—which appeared in March 2010 in the *Journal of Human Evolution*. Written by early primate experts Blythe Williams, Richard Kay, Christopher Kirk, and Callum Ross, the paper reaffirmed that *Darwinius* had far more in common with lemurs than monkeys.

[4] My miniscule contributions to the public discussion about *Darwinius* included several blog posts, two pieces in the *Times* of London, and two appearances on the BBC Radio 4 show *Material World*. I hope they helped at least a few people understand that Ida was not all she was cracked up to be.

[5] Though, as Adams so astutely pointed out, perhaps we should be more concerned with asking the right questions first.

[6] If something that was imaginable didn't exist, it was argued, the thing either had not been found or possessed some inherent contradiction that barred the actual possibility of its existence.

[7] There are some notable exceptions, of course, such as the Blob (a living, oozing metaphor for communism) and the parasitic stars of the *Alien* series, but for every such creature there seems to be a score of big-brained humanoids.

[8] In the interest of fairness, however, it should be noted that Dawkins does consider the fossil record as evidence for evolution in his latest book, *The Greatest Show on Earth*.

[9] The solution to this problem would only become possible with the invention of radiometric dating. After the discovery of radioactivity in the late nineteenth century it became apparent that isotopes of certain elements contained within certain rocks, such as uranium 238 and potassium 40, break down or decay into isotopes of other elements at constant rates over time (lead 206 and argon 40, respectively, in the case of the two mentioned here). By observing the rate at which the original radioactive material decayed into the daughter isotopes scientists could then look at the ratio of "parent" to "daughter" material in rock and determine how long it would have taken to produce that amount of daughter material, thus yielding the rock's age.

Unfortunately, most fossils are contained within sedimentary rocks that do not contain radioactive materials (with the exception of strata made of ash from volcanoes), but many times sedimentary strata are embedded between lava flows or rocks with radioactive materials. By dating the rocks above and below the fossil layers, then, geologists can establish a time frame. In this way "relative dating" based upon fossils and "absolute" dating work together.

[10] Charles Darwin would later be a student of Jameson, but the budding sixteen-year-old naturalist found Jameson to be a rather dull teacher.

[11] Jameson's influence on this point has been known by historians of scientists for some time, but the myth that Cuvier was a religious fundamentalist who set out to prove a global Flood remains widespread. I heard it myself several times during my education. The fact that Cuvier argued forcibly against the evolutionary ideas of Lamarck and Geoffroy St. Hilaire bolstered this misunderstanding, for surely anyone who opposed evolution at the time must have done so on religious grounds (or so the myth goes).

[12] Nor was Buckland especially credulous. He had gained fame for his ability to debunk fantastic claims. Perhaps most notably, during his honeymoon with his new wife in Italy in 1825, Buckland visited the shrine of Santa Rosalia, a Catholic saint whose bones were said to have eradicated the plague from the town of Palermo in 1624. When Buckland saw the bones, however, his expertise in comparative anatomy allowed him to immediately recognize them as goat, not human, bones. The attending priests replied that only the truly devout (i.e., Catholics) could see the bones, and since that time no one has been allowed to see "Rosalia."

[13] Today we know that Buckland was indeed studying evidence of flooding, just not of the sort he imagined. The gravels, loams, and clays were deposited by melting ice sheets during glacial cycles over the previous two and a half million years. When the climate warmed, the ice sheets melted, flooding the surrounding area and dropping mixtures of geological bric-a-brac that had been carried along from places far distant.

[14] Traditionally Cuvier and Buckland's view has been known as "catastrophism" and the opposing view as "uniformitarianism." I have eschewed these labels here as I believe that they are more often abused these days than used properly. Even though Buckland and Cuvier spoke of quick revolutions they also relied on geological processes now in action to understand less drastic changes. Likewise, processes now operating cannot entirely account for geological events of the past, particularly events like asteroid strikes. Modern geology is thus a combination of both systems and it is not profitable to identify one side as the victor despite a textbook preference for the label uniformitarianism.

[15] There is so much to discuss about Charles Darwin that it cannot all fit in one chapter. Some readers will surely say, "But you forgot . . ." by the time they get to the end of it. I can only offer my apologies that I did not have more space. Given the interests of this book I have decided to focus on the importance of geology and paleontology to Darwin's work. There is no shortage of biographies on Darwin for those who desire a more comprehensive account, among the best being Adrian Desmond and James Moore's *Darwin: The Life of a Tormented Evolutionist* and Janet Browne's two-volume study, *Darwin: Voyaging* and *Darwin: The Power of Place*.

[16] My favorite tale is that of the beaver. The testicles of male beavers were useful as medicine, the *Physiologus* says, and so male beavers were often hunted for this resource. Rather than let themselves be slaughtered, however, the beavers would castrate themselves and throw their genitals to the hunter so that they might escape with their lives. Should they be harassed by another hunter later they would simply roll over to show that they had already been snipped. And the moral of this tale of self-mutilation? That we must likewise excise sin from our lives and cast it before the devil so that he will leave us alone. Never mind that the testicles of beavers are actu-

ally held inside their bodies, making them incapable of exemplifying such a lesson.

[17] Two days before the publication of *On the Origin of Species* Darwin would write to his neighbor John Lubbock that "I do not think I hardly ever admired a book more than Paley's *Natural Theology*: I could almost formerly have said it by heart."

[18] By this time Darwin had become a well-respected young naturalist, so his father organized investments so that Darwin would be financially supported.

[19] Natural selection was the core mechanism in Darwin's evolutionary framework, but it was not the only one. He believed that there were other supplementary mechanisms that also contributed to species change, and he was often frustrated when critics asserted that his vision of evolution was based on natural selection and nothing else.

[20] Owen had a habit of changing species and genus names if he thought the original title was somehow inaccurate or not evocative enough. In this particular case the name change was not warranted, and today the lungfish from Africa is known by the name *Protopterus*.

[21] It should be noted, however, that not all tetrapods develop fingers from this arc. There are other patterns of finger development, usually associated with a reduced number of fingers, but the mode of digit formation as proposed by Shubin and Alberch is the most relevant for the discussion of early tetrapods.

[22] There is still some debate whether the hands and feet of tetrapods were entirely novel features. Some scientists have argued that that they are modified versions of structures seen in the ancestral fish, while others have said that the differences between zebrafish and tetrapod embryos show that hands and feet were entirely unique to tetrapods. The problem with this argument is that zebrafish are not particularly closely related to tetrapods and might not be very informative on this point. Research and debate will continue, but the basic developmental patterns outlined in this chapter have been confirmed by multiple lines of evidence.

[23] And for those who are about to object "We don't have tail bones!" I must point out that you are probably sitting on them right now. The coccyx in our skeleton is the remnant of a tail possessed in our monkey ancestors and remains an attachment point for sinew and muscle.

[24] In January 2010 a team of scientists led by Ahlberg described what appeared to be tetrapod trackways over 10 million years older than *Tiktaalik* in the journal *Nature*. If these traces were made by tetrapods then the evolution of the first vertebrates with limbs was far more complex than we presently realize, but whether the "tracks" were made by tetrapods at all is still controversial. Many tracks attributed to tetrapods have later turned out to be traces left by invertebrates, from giant centipedelike arthropods to sea stars that left fingerlike impressions, so further research will be required to determine what animals left the traces.

[25] The historian Adrienne Mayor has done more than anyone else to elucidate the connections between fossils and mythology. Her books *The First Fossil Hunters* and *Fossil Legends of the First Americans* are essential reading on geomythology.

[26] The association of the Lenape with Europeans led to the near-destruction of the tribe. Disease, confusion over property rights, and reliance on trade with the Europeans all contributed to the downfall of the Lenape, but there are still Native Americans today who can trace their ancestry back to this tribe.

[27] Deane did not publish his first paper on the footprints until 1843, eight years after he became acquainted with them. The fact that Hitchcock published earlier and more frequently than Deane, however, led to a public row over who had priority. Hitchcock tried to strike a conciliatory note, while maintaining that he had launched a new branch of science, but Deane was not satisfied. Hitchcock ultimately won most of the recognition, but Deane's contributions should not be forgotten.

[28] The more famous poet Henry Wadsworth Longfellow was also so inspired by

the tracks. While he did not devote an entire poem to them, they are mentioned in both "To the Driving Cloud" and "A Psalm of Life." These were the "footprints on the sands of time."

[29] Owen did ascribe it to a new species on the basis that the feathers with the skeleton looked slightly different than the feather von Meyer described. It was not unreasonable to think that there could have been more than one bird present at the same place at the same time, but even though there is still controversy today most paleontologists ascribe all the known specimens to the species established by von Meyer, *Archaeopteryx lithographica*.

[30] Phillips had been suspicious that parts of the *Megalosaurus* were out of order prior to Huxley's visit, but he had not yet completed his research on the problem. This story, recounted by Huxley, also runs counter to the apocryphal story that Huxley first recognized the resemblance between birds and dinosaurs while carving a Christmas goose. I do not know the origins of this tale, but I have found no evidence for its veracity. Even so, many paleontologists do subject their friends and relatives to anatomy lectures every Thanksgiving and Christmas.

Additionally, a recent reanalysis found that many of the fossils attributed to *Megalosaurus* actually belong to other dinosaurs, meaning that the original jaw fragment Buckland described is all we presently know of it.

[31] Owen was able to give form to his view through the artistic talent of Benjamin Waterhouse Hawkins, who was commissioned to create life-size sculptures of the dinosaurs (and other prehistoric monstrosities) for the Crystal Palace exhibition of 1851 in Hyde Park, London. Hawkins's sculptures were later moved to Sydenham Hill, where they still can be seen today.

[32] Of *Hesperornis* the ornithologist William Beebe would later write, "When in the depth of the winter, a full hundred miles from the nearest land, one sees a loon in the path of the steamer, listens to its weird, maniacal laughter, and sees it slowly sink downward through the green waters, it truly seems a hint of the bird-life of long-past ages."

[33] Even creationists offered ideas for the point of origin of birds. In an article published in 1897, W. T. Freeman wrote, "I suggest that in the earlier days there were ill-developed, low-typed, wallowing birds, also some highly developed reptiles. Perverted sexual instinct exists now, why not then, and as a result of this, why has not the archaeopteryx been an anomalous false hybrid that has been incapable, like other mongrels, of reproducing its kind?"

[34] Although he doubted its ability to fly, Beebe thought that *Archaeopteryx* might have been able to sing (or at least squawk). A naturalist walking through a Jurassic forest might have heard "an archaic attempt at song—a lizard's croak touched with the first harmony, which was to echo through all the ages to follow." This same idea was expounded several years earlier in Eden Phillpotts's fictional work *Fancy Free*. In the work, a fictional clergyman recounts a Mesozoic safari, during which he spies an *Archaeopteryx*, "He was the very first thing of his kind that Nature had managed; naturally he could conceive of nothing finer than his primitive self and preposterous voice. He gurgled and hissed, and squeaked, and even tried to trill."

[35] This is an oversimplification of Dollo's Law, stated here as it best reflects Heilmann's reasoning. As it is presently understood, Dollo's Law has more to do with the closing of certain evolutionary pathways as creatures change. Snakes, for example, lost their legs through evolution and cannot revert to an ancestral state. Descendants of living snakes could evolve legs, but they would be entirely new structures and not duplicates of the ones their distant ancestors lost.

[36] In 2007 paleontologists announced that they had finally found evidence of group behavior in these kinds of dinosaurs. A trackway from China not only recorded the tracks of *Deinonychus* relatives, confirming that they held their large claws off the

ground, but the trackway also preserved the movements of several dinosaurs together. The intricacies of the preservation and spacing of the tracks supported the idea that these dinosaurs were moving together, and it seems that at least some dinosaurs like *Deinonychus* were gregarious.

[37] Even this is an oversimplification. Some animals, like small bats and birds, have high constant body temperatures for part of the day or year but have their body temperatures affected by the surrounding environment at other times. Others, like some fish, have body temperatures that fluctuate with the environment but are maintained several degrees higher than the surrounding water.

[38] This was later reinforced by the discovery of another *Archaeopteryx* without clear feather impressions that had been labeled *Compsognathus*. What so many naturalists said was right; without feathers *Archaeopteryx* looked just like a dinosaur.

[39] Artist Gregory Paul illustrated what some of these downy dinosaurs might have looked like in his 1988 book *Predatory Dinosaurs of the World*.

[40] Other groups of predatory dinosaurs, represented by theropods, such as *Allosaurus*, *Spinosaurus*, and *Carnotaurus*, were not coelurosaurs, and to date there is no evidence that they had feathers.

[41] This is no more fantastic than suggesting that our own prehistoric ancestors were covered in hair, based upon their relationship to us.

[42] There are a number of ways to distinguish the ornithischians from the saurischians, but the easiest is to look at their hips. The ornithischians have a backward-pointed process called the pubis, while the same process in many saurischians is oriented forward (though it was secondarily rotated to point backwards in dinosaurs closely related to birds). Huxley was wrong when he thought the hips of *Hypsilophodon* represented all dinosaurs. We now know it was a type of ornithischian, unrelated to the earliest birds.

[43] Some marine reptiles, like the ichthyosaurs, had only one temporal fenestra. They differed from synapsids, however, in that they had evolved from ancestors with two such openings and the one that was retained was higher up on the skull.

[44] Today the term "pelycosaur" has fallen out of fashion because it is thought to be paraphyletic, or a group that does not contain all the descendants of a common ancestor (in this case, the therapsids). The name is retained here, however, because it allows creatures like *Dimetrodon* to be distinguished from other synapsids, and the connection between the pelycosaurs and the earliest therapsids is explained.

[45] An exception to this may be a potential early therapsid fossil from the Lower Permian called *Tetraceratops*. More complete remains will be required to determine its relationships, but if it is an early therapsid, as some paleontologists have proposed, it may indicate that the early therapsid form existed for millions of years before the adaptive radiation at about 267 million years ago. At present, though, it appears that therapsids diversified quickly after the first members of the group evolved, fitting a more punctuated evolutionary model than a "gradual" one.

[46] Interestingly, the separation of ear bones from the lower jaw of synapsids may have occurred more than once. In 2005 a team led by Thomas Rich described the jaw of an early platypus relative called *Teinolophos* that lived after the split between monotremes and other mammals but still had ear bones connected to its lower jaw. This suggests that the separation of the ear bones from the lower jaw happened once among monotremes and again in the lineage leading to the other two groups of mammals.

[47] The levels at which natural selection might work have been hotly debated in recent years. Some scientists, such as Richard Dawkins, have argued that natural selection primarily works at the level of the gene, thus rendering organisms only as gene-propagation vehicles. Scientists such as Stephen Jay Gould, however, have argued that natural selection can act on a variety of hierarchical levels, from genes to

individuals to populations to species and maybe even entire evolutionary groups. I am more inclined to agree with the latter camp on this point, and the details of mass extinctions and their aftermath might be able to tell us how natural selection acts at a level above that of individual organisms.

[48] The intense eruptions of the Deccan Traps volcanoes have also been proposed as extinction triggers, but they do not fit the pattern of extinction as presently understood. Even so, the cause of the end-Cretaceous mass extinction is still being debated, and future discoveries will help paleontologists better understand what happened.

[49] H. N. Hutchinson also mourned the loss of so many specimens in his book *Extinct Monsters and Creatures of Other Days*: "How little [do the farmers] know that hundreds of museum curators all over the world would be only too glad to procure some of this 'rubbish'!"

[50] Like many fossils in German museums, Koch's original *Hydrarchos* was destroyed in an Allied bombing raid on Berlin in 1945.

[51] There are some notable exceptions to this. Extinct archosaurs such as *Tyrannosaurus* and *Malawisuchus* had heterodont teeth, and some mammals, like dolphins, have homodont teeth (or teeth that are the same throughout the jaw).

[52] Agassiz later changed his mind when he saw them himself and said that the jaws came from mammals.

[53] Interestingly, remains of *Pakicetus* had been described just the year before as a new species of *Protocetus* called *Protocetus attocki*. The fact that those particular fossils indicated a second species of *Pakicetus* could not have been known until more complete remains of the genus were discovered and compared.

[54] Sadly, the Yangtze River dolphin, also known as the Baiji, was declared extinct as this book was being prepared. There may be a few individuals left, but the population is so small as to no longer be viable. The reason for the extinction of the Baiji is directly attributable to human activities, particularly hunting during the humanitarian disaster known as the Great Leap Forward, pollution of the river, and habitat reduction as a consequence of the construction of the Three Gorges Dam. All of the other freshwater dolphins in the world are endangered as well.

[55] Think of doing the breaststroke in a pool. Half the motion is the actual stroke that propels you forward and the other half is a moment of little forward movement in which you bring your limbs back into position for the next stroke.

[56] One hypothesis is that the legs might have been used in nuptial encounters, but just because the structures exist does not mean we need to invent things for them to do. It may be that the small hind limbs were not used for anything; all the more reason for them to be evolved away.

[57] The reasons seals and sea lions (or pinnipeds) swim using their fore-and hind-limbs represent an alternate evolutionary path for aquatically adapted mammals. In April 2009 Natalia Rybczynski, Mary Dawson, and Richard Tedford announced the discovery of *Puijila darwini*, an otterlike relative of seals and sea lions. A transitional fossil connecting these marine mammals with their terrestrial ancestors, *Puijila* had very robust arm and leg bones, suggesting that it primarily swam by using its webbed hands and feet. This mode of swimming likely had as much influence over the evolution of pinnipeds as undulation had among early whales.

[58] Other extinct aquatic mammals overcame the buoyancy problem in different ways. During the Miocene and Pliocene there was a genus of giant sloths (*Thalassocnus*) that lived along the coast of what is now Peru. They appear to have been at least semi-aquatic, swimming into the shallows to eat algae and other soft plants. They show no sign of using gastroliths or developing bone ballast. Instead they may have held themselves beneath the water by grabbing onto rocks with their massive claws.

[59] Buffon chose iron as the best model metal for Earth, and through his experiments he made the more controversial claim that the earth was much older than the

6,000 and some odd years that Biblical chronologists had frequently come up with.

[60] The moose, said to be about seven feet tall, arrived in Paris in 1787, but by then Buffon had left Paris due to his failing health. The haste with which the carcass was sent caused it to deteriorate rapidly, and despite Jefferson's wish to see the specimen properly stuffed and displayed in the city's great natural history museum, many Parisians were probably hoping that Jefferson would send the smelly thing away.

[61] At the time Virginia encompassed a much larger area than it does today. "Big Bone Lick" is currently within the borders of the state of Kentucky.

[62] The imposing teeth and ornery manner of hippos made many early explorers wonder whether they were carnivorous. It would later become known that they are primarily herbivorous, but they will not turn up their noses at meat if an opportunity to scavenge presents itself. In fact, in 2004 it was reported that hippos had become so overcrowded in parks of Uganda that they began to cannibalize each other at a much higher rate, perhaps hastening the spread of a deadly anthrax outbreak that was working its way through the population at that time.

[63] Some, however, thought that the bones belonged to giants that were human in proportion. For a time this was the opinion of Yale president Ezra Stiles, but Jefferson could not accept this conclusion based upon the presence of tusks with the other large bones. After corresponding with Jefferson, Stiles changed his mind.

[64] Thomas Jefferson described the claws of a similar animal in 1797, which he called *Megalonyx*, but with only the fossil talons to work with he thought it was a kind of enormous lion. The fossils would not be officially named in Jefferson's honor until 1825, and by this time it was known that the claws had belonged to a giant ground sloth.

[65] The hypothesis that *Amebelodon* and *Platybelodon* used their jaws exclusively to scoop up water plants would later be refuted. These animals lived in terrestrial habitats, and the wear on the enamel of their teeth is inconsistent with the kind of habits paleontologists like H. F. Osborn envisioned. They may have used their teeth and jaws as scoops sometimes, but they also used them to strip bark off trees, saw through tough plants, and gather plant food through other methods.

[66] In his contribution to the *Bridgewater Treatises* the English naturalist William Buckland would later speculate that *Deinotherium* used its lower jaw tusks to anchor itself to the riverbank while sleeping, but there is no evidence that it did so.

[67] These names can be a little confusing as they are often derived from common roots and only have different suffixes. Generally, though, any fossil vertebrate group ending in "-formes" contains a wider variety of creatures, or archaic branches closely related to a more specific, specialized group. This latter group is often distinguished by having the suffix "-morpha." Another example would be the archosaurs close to the ancestry of dinosaurs. The dinosauriformes were a diverse group of early, dinosaurlike animals, while the dinosauramorpha (nested within the dinosauriformes) contains dinosaurs and their closest, non-dinosaurian relatives.

[68] Reptiles do not have this problem, as they constantly replace teeth throughout their lives, but mammals only have two sets: milk teeth that are shed and adult teeth that must last mature animals their entire lives.

[69] I have decided to leave out a discussion of the Pleistocene impact hypothesis for now, as we cannot yet be certain that such an event actually occurred or that it caused an extinction that, by all appearances, was already in progress at the time. Discussing the evidence so far would require an undue amount of speculation on my part. For now I am content to wait for more complete evidence.

[70] In fact, in 2009 a study based upon genetic material preserved in the permafrost of Alaska suggested that there were mammoths living in that location until about 7,000 years ago.

[71] And this is to say nothing of more practical concerns, such as inadvertently in-

troducing new diseases to animals in North America, the threat that some introduced animals might escape, and the cost to maintain the fencing around hypothetical "Pleistocene parks." Even if agreement could be reached about the effects of such a "rewilding" attempt, the practical, economic, and societal concerns might stall such projects. Then again, there are already many-acre private exotic game ranches and elephant refuges in the United States, so perhaps these problems could be overcome with enough support. Whether it is a project worth undertaking, however, is another question.

[72] Linnaeus also described two other species of *Homo*. One, *Homo troglodytes*, was said to be a nocturnal species that only communicated in hisses; while the other, *Homo caudatus*, was a humanlike inhabitant of the Antarctic that sported a tail. The existence of these species was based on hearsay, but absence of proof was not considered proof of absence. Perhaps there were other humanoids waiting to be found.

[73] The school of biblical analysis called higher criticism, at least partly inspired by Peyrère's work, was also important to this change. These scholars strove to understand religious documents by finding independent evidence to confirm (or sometimes refute) certain interpretations of scripture. Their conclusions that did not conform to traditional beliefs (like the divine creation of humanity) were decried as heretical, but the controversy higher criticism stirred reaffirmed that Genesis did not have to be read as a history book.

[74] By this time, the Indonesian apes we know as orangutans today had already been placed in their own genus, *Pongo*. Gorillas (the "Pongo" of Battell) and chimpanzees had also been described, although the species of chimpanzee called the bonobo would not be identified until the 1920s. Historically, though, the same names were often used to refer to several different kinds of primates, and this created a confusing morass of names. Chimpanzees, for instance, were variously referred to as apes, baboons, mandrills, jockos, and orang-outangs, among others.

[75] A later investigation of the skeleton by W. J. Sollas, Buckland's successor at Oxford, determined that the woman was really a young man. The pelvis is the most useful bone in determining the sex of an individual, as this bone contains several features (like the width of the sciatic notches on the sides of the hips) that allow osteologists to tell the difference between males and females.

[76] Similar skulls were discovered in Gibraltar in 1829 and in Belgium in 1848, but their significance was overlooked until well after the 1856 discovery was made.

[77] The exact reason why Dubois cut off access to the fossils is unknown. The resentment his actions caused among anthropologists led to wild accusations about getting even with his critics, religious conversion, and other stories. It is more likely attributable to several factors, including Dubois' difficulty accepting criticism, his annoyance at interruptions when people wanted to see the fossils, and other people publishing descriptive papers about "his" fossils. Even these considerations might not reveal the entire story. In an interview with the anthropologist W. K. Gregory printed in *Popular Science* in 1931, Gregory said that during this time Dubois was carefully removing the sediment from beneath the skull in order to make a plaster cast of it to show what the brain might have looked like. Even if this is true, though, there were likely other reasons why Dubois was so possessive of the bones.

[78] But their importance to Chinese culture would endure. In communist China, especially, the bones of "Peking Man" were seen as exhibiting the virtue of the worker; it had been through the use of stone tools that humans had pushed their evolution forward. A full study of the importance of these fossils to Chinese culture can be found in Sigrid Schmalzer's *The People's Peking Man*.

[79] Dart almost orphaned his petrified child when he forgot it in the taxi he took back to his hotel. Fortunately, he was able to recover it.

[80] The conference marked a turning point for Leakey. He had been shamed by

two scientific mistakes: the identification of a recent burial as the oldest *Homo sapiens* ("Oldoway Man") and misidentifying geologically recent *Homo sapiens* as an even older ancestor ("*Homo kanamensis*"). He had a reputation as a sloppy field worker, and he was engulfed in personal scandal when he left his wife with child for a young illustrator named Mary Nicol (who later married Leakey). The 1947 conference, in which Leakey could show off on his home turf, did much to rehabilitate his career.

[81] The story of the Piltdown Man forgery requires more space and attention than I can provide here. For a full account, see *Piltdown: A Scientific Forgery* by Frank Spencer.

[82] But who could have perpetrated such an act of scientific malfeasance? Everyone was a suspect. That Dawson was involved was nearly a certainty, but author Arthur Conan Doyle, paleontologist Teilhard de Chardin, anthropologist Arthur Keith, and several others were not beyond scrutiny as potential accomplices or masterminds. With Dawson dead and no further clues, however, the motives and the identity of any other hoaxers remain mysterious.

[83] Leakey also sent Dian Fossey to study mountain gorillas in Rwanda in 1967 and Biruté Galdikas to study orangutans in Borneo in 1971. These are the most famous women to go on these excursions, but they were not the only ones. Less well known are Rosalie Osborn, who went to study the gorillas in October 1956, and Jill Donisthorpe, who continued the gorilla observations from January 1957 through September of that year. Both had not learned much more than the little that was already known about gorillas, and this general lack of information led George Schaller and his graduate advisor to mount their own scientific expedition in 1958.

[84] In 2008, researchers reported that bone tools associated with *Paranthropus* were found at the 1.5–2-million-year-old site at Drimolen in South Africa. These were not weapons, but tools that appear to have been used for digging, perhaps to get into termite mounds or uncover tubers in the ground.

[85] As strange as it may seem, the predators that ate our prehistoric relatives often preserved their bones. The "First Family" of *Australopithecus afarensis* may have been a saber-toothed cat kill; damage to the skull of the "Taung Child" shows that it was a meal for an eagle; the skullcaps of *Homo erectus* at Dragon Bone Hill are the tablescraps of giant hyenas; and *Homo habilis* foot bones from Olduvai Gorge show crocodile toothmarks. In fact, in February 2010 a team of scientists led by Chris Brochu discovered a new fossil crocodile from Olduvai Gorge which could have been capable of doing the damage seen on the *Homo habilis* bones, and so the scientists named it *Crocodylus anthropophagus* (literally, the "man-eating crocodile").

[86] The parts of the study based on social behaviors was led by Owen Lovejoy, a locomotion specialist who had previously claimed that the evolution of bipedalism in humans was related to males provisioning females with food to gain sex, thus leading to monogamy, decreased sexual dimorphism, etc. Lovejoy essentially repackaged these assertions, which had generally been panned by other anthropologists, for *Ardipithecus ramidus*, and the inferences about social structure are the weakest parts of the new study.

[87] In turn this would suggest that gorillas and chimpanzees do not represent the anatomy of our last common ancestors, but instead are specialized apes that have evolved in different ways (such as independently acquiring a knuckle-walking posture). They are still useful for thinking about our own evolution, but we cannot take them to be what our earliest ape ancestors were like.

[88] In April 2010, paleoanthropologists from South Africa announced the discovery of a new species of australopithecine they named *Australopithecus sediba*. Although proposed as a possible ancestor of our own genus, *Homo*, this 1.7-million-year-old hominin is a little too young and a little too different to be one of our direct ancestors. Instead, it seems that *A. sediba* was a unique, specialized species of human

that lived alongside some of the earliest members of our own genus.

⁸⁹ There is some dispute about these latter skulls, specifically whether they belong to *Homo erectus* or a closely related species called *Homo georgicus*. Further study will hopefully clarify this issue.

⁹⁰ And there may have even been another species of human alive during this time. On March 24, 2010, it was announced in the journal *Nature* that mitochondrial DNA recovered from a 40,000-year-old finger bone found in Siberia's Altai Mountains differed strongly from both archaic human and Neanderthal DNA. What this means is debatable. The disparity between the DNA samples may signal the existence of an unknown species of human, a *Homo sapiens*-Neanderthal hybrid, or a population of Neanderthals that had been genetically isolated from their European relatives (which have formed the base of what we know about Neanderthal DNA). Which of these hypotheses, if any, is correct will require further evidence to determine.

⁹¹ Several years later, paleontologist Dougal Dixon presented a more comprehensive vision of the dinosaurs' "alternative evolution" in his book *The New Dinosaurs*. (Along with his book *After Man*, it is one of the most ingenious collections of speculative biology ever written.) Carl Sagan also speculated on this point in his book on the origins of intelligence, *The Dragons of Eden*, in which he also wondered if a maniraptoran dinosaur might evolve into a highly intelligent species comparable to our own.

References

INTRODUCTION: MISSING LINKS

p. 9   "My curiosity was initially piqued…" – Churcher, Sharon. 2009. Is David Attenborough Set to Reveal the Missing Link in Human Evolution? *Mail Online*, May 9, Science & Tech Section. http://www.dailymail.co.uk/sciencetech/article-1179926/Is-David-Attenborough-set-reveal-Missing-Link-human-evolution.html

p. 10   "A May 15, 2009 piece by the *Wall Street Journal*…" – Naik, Gautam. 2009. Fossil Discovery Is Heralded. *Wall Street Journal*, Mary 15, World section. http://online.wsj.com/article/SB124235632936122739.html

p. 10   "When the paper was finally released…" – Franzen, J. L., P. D. Gingerich, J. Habersetzer, J. H. Hurum, J. von Koenigswald, and B. H. Smith. 2009. Complete Primate Skeleton From the Early Eocene of Messel in Germany: Morphology and Paleobiology. *PLoS One 4*, no. 5 e5723 (May 19) http://www.plosone.org/article/info:doi/10.1371/journal.pone.0005723

p. 11   "He proclaimed that Darwinius was…" – University of Oslo Natural History Museum. Press Release – The Link. http://www.revealingthelink.com/more-about-ida/resources/press_release.pdf

p. 11   "…lead author Jens Franzen stated that the effect of their research…" – Leake, Jonathan, and John Harlow. 2009. Origin of the Specious. *Times Online*, May 24, Science section. http://www.timesonline.co.uk/tol/news/science/article6350095.ece

p. 11   "*New York Times* journalist Tim Arango…" – Arango, Tim. 2009. Seeking a Missing Link, and a Mass Audience. *New York Times*, May 18, Business section. http://www.nytimes.com/2009/05/19/business/media/19fossil.html

p. 11   "Any pop band is doing the same thing…" – Arango, Tim. 2009. Seeking a Missing Link, and a Mass Audience. *New York Times*, May 18, Business section. http://www.nytimes.com/2009/05/19/business/media/19fossil.html

p. 11   "When the fossil pit in Messel, Germany…" – Tudge, Colin, and Josh Young. 2009. *The Link: Uncovering Our Earliest Ancestor*. New York: Little Brown and Company. pp. 8-36.

p. 12   "Hurum also had bigger things in mind…" – Coghlan, Andy. 2009. Fossil of 'Ultimate Predator' Unearthed in Arctic. *New Scientist*, March 17. http://www.newscientist.com/article/dn16785-fossil-of-ultimate-predator-unearthed-in-arctic.html

p. 13   "Team member Philip Gingerich would later lament…" – Dayton, Leigh. 2009. Scientists Divided on Ida as Missing Link. *The Australian*, May 20, The Nation section.

p. 13 "According to one of the reviewers..." – Randerson, James. 2009. Is the Ida Fossil a Missing Evolutionary Link? *The Guardian*, May 19, Science section. http://www.guardian.co.uk/science/2009/may/19/fossil-ida-missing-link

p. 14 "In 2001, five years prior to the sale of *Darwinius*..." – Gibbons, Ann. 2009. New Primate Fossil Poses Further Challenge to Ida. *ScienceNow*, October 21. http://news.sciencemag.org/sciencenow/2009/10/21-02.html

p. 14 "The Fayum team spent years piecing together..." – Seiffert, E., J.M.G. Perry, E.L. Simons, and D.M. Boyer. 2009. Convergent Evolution of Anthropoid-like Adaptations in Eocene Adapiform Primates. *Nature* 461: 1118-1121.

p. 17 "Distancing himself from the headline-making claims..." – TheScientist.com blog, http://www.the-scientist.com/blog/display/56110/

p. 17 "Gingerich was similarly unimpressed..." – Keim, Brandon. 2009. Bone crunching debunks 'first monkey' Ida fossil hype. *Wired Science*, October 21. http://www.wired.com/wiredscience/2009/10/reconfiguring-ida/

p. 18 "If a landed proprietor is asked to produce..." – Huxley, T.H. 1868. On the animals which are most nearly intermediate between birds and reptiles. *Geological Magazine* v: 357-365.

p. 19 "This was the Great Chain of Being..." – Lovejoy, Arthur. 1964. The Great Chain of Being. Cambridge: Harvard University Press.

p. 20 "In his 1870 address as president of London's Geological Society" – Huxley, T. H. 1877. On the study of biology. *Nature* XV: 219-224.

p. 21 "In the introductory chapters of his 2004 tome *The Ancestor's Tale*..." Dawkins, Richard. 2005. The Ancestor's Tale. Boston: Houghton Mifflin. pp. 13.

p. 22 "This precipitated a 1980 conference in Chicago..." – Lewin, Roger. 1980. Evolutionary theory under fire. *Science* 21: 883-887.

THE LIVING ROCK

p. 24 "In October 1666 a French fishing boat..." – Cutler, Alan. 2004. The Seashell on the Mountaintop. New York: The Penguin Group.

p. 25 "The earth, through some kind of plastic force..." – Rudwick, Martin. 1985. The Meaning of Fossils. New York: Science History Publications, USA. pp. 56.

p. 27 "Steno presented his thoughts..." – Winter, John Garret. 1916. The Prodromus of Nicolaus Steno's Dissertation. New York: The Macmillan Company.

p. 28 "In a scheme presented by English naturalist John Woodwards..." – Woodwards, John. 1723. An Essay Towards a Natural History of the Earth (Third Edition). London.

p. 29 "Hutton was an eccentric polymath..." – Repcheck, Jack. 2003. The Man Who Found Time. Philadelphia: Perseus Books.

p. 30 "Two factors impeded the acceptance of Hutton's view..." – Gould, Stephen Jay. 1996. Time's Arrow, Time's Cycle. Cambridge: Harvard University Press. pp. 61-97.

p. 30   "During the late eighteenth century..." – Rudwick, Martin. 1997. Georges Cuvier. Chicago: University of Chicago Press.

p. 32   "William Smith, an English geologist..." – Winchester, Simon. 2001. The Map That Changed the World. New York: HarperCollins.

p. 32   "Like Smith, Cuvier also recognized..." - "The earth, through some kind of plastic force..." – Rudwick, Martin. 1985. The Meaning of Fossils. New York: Science History Publications, USA. pp. 127-131.

p. 33   "Cuvier most explicitly outlined the role of catastrophes..." – Rudwick, Martin. 1997. Georges Cuvier. Chicago: University of Chicago Press. pp. 248.

p. 33   "William Buckland was a..." – Rudwick, Martin. 2009. Worlds Before Adam. Chicago: University of Chicago Press.

p. 34   "It was a simple discovery..." – Davy, John. 1840. The Collected Works of Sir Humphry Davy (Vol. VII). London: Smith, Edgar and Co. Cornhill. pp. 40.

p. 34   "Encouraged by his peers..." – Buckland, William. 1823. *Reliquiae diluvianae, or, Observations on the Organic Remains attesting the Action of a Universal Deluge.* London: John Murray.

p. 36   "Every geological process had its counterpart..." – Lyell, Charles. 1837. Principles of Geology. Philadelphia: James Kay, Jun & Brother. pp. 81.

p. 37   "Hutton himself had considered this..." – Pearson, Paul N. 2003. Excerpt from An Investigation of the Principles of Knowledge. *Nature* 425: 665.

p. 37   "The most prominent exponent of this view..." – Packard, Alpheus. 1901. Lamarck, the Founder of Evolution. New York: Longmans, Green and Co.

p. 39   "During such a time, Lyell proposed..." - "Every geological process had its counterpart..." – Lyell, Charles. 1837. Principles of Geology. Philadelphia: James Kay, Jun & Brother. pp. 131.

MOVING MOUNTAINS

p. 42   "In the evening hours of January 19, 1836..." – The Complete Work of Charles Darwin Online. Darwin's *Beagle Diary* (1831-1836) http://darwin-online .org.uk/content/frameset?itemID=EHBeagleDiary&viewtype=text&pageseq=1

p. 43   "Given that Darwin's father was a physician..." – Desmond, Adrian, and James Moore. 1991. Darwin: The Life of a Tormented Evolutionist. New York: Warner Books.

p. 45   "According to Paley the whole of nature..." – Paley, William. 2006. Natural Theology. New York: Oxford University Press.

p. 47   "It was not an easy journey." – Nichols, Peter. 2003. Evolution's Captain. New York: HarperCollins.

p. 48   "Though not a paleontologist..." – Brinkman, Paul. 2009. Charles Darwin's Beagle Voyage, Fossil Vertebrate Succession, and "The Gradual Birth & Death of Species." Journal of the History of Biology (June 16), http://www.springer-link.com/content/k6304m6w42351411/?p=965a149cbcc44357a349140f928a a3d5&pi=1

p. 49   "In a field notebook entry made in February of 1835..." – The Complete

Work of Charles Darwin Online. The position of the bones of Mastodon (?) at Port St Julian is of interest. http://darwin-online.org.uk/content/frameset ?itemID=CUL-DAR42.97-99&viewtypo=text&pageseq=1

p. 52   "As he began to jot down notes..." - The Complete Work of Charles Darwin Online. Darwin's Notebooks on Transmutation of Species. http://darwin-online.org.uk/content/frameset?itemID=F1574a&viewtype=text&pageseq=1

p. 52   "In the spring of 1837..." - Desmond, Adrian, and James Moore. 1991.Darwin: The Life of a Tormented Evolutionist. New York: Warner Books. pp. 222.

p. 55   "William Buckland, the Oxford geologist..." – Buckland, William. 1836. Geology and Mineralogy Considered With Reference to Natural Theology. London: William Pickering.

p. 56   "This view was formally articulated by Charles Babbage..." – Babbage, Charles. The Ninth Bridgewater Treatise. London: John Murray.

p. 57   "Entitled *Vestiges of the Natural History of Creation*..." – Chambers, Robert. 1844. Vestiges of the Natural History of Creation. London: John Churchill.

p. 58   "A reply published the following year..." – Bosanquet, S.R. 1845. Vestiges of the Natural History of Creation: Its Argument Examined and Exposed (Second edition). London: John Hatchard and Son.

p. 59   "Hooker broached the subject..." – Darwin Correspondence Project. Letter 804. http://www.darwinproject.ac.uk/entry-804

p. 59   "Darwin replied..." – Darwin Correspondence Project. Letter 814. http://www.darwinproject.ac.uk/entry-814

p. 59   "While Darwin settled down to a quiet life..." – Carroll, Sean. 2009. Remarkable Creatures. Orlando: Houghton Mifflin. pp. 47-59.

p. 59   "This sounded very familiar..." – Darwin Correspondence Project. Letter 1792. http://www.darwinproject.ac.uk/entry-1792

p. 60   "On May 1, 1857..." – Darwin Correspondence Project. Letter 2086. http://www.darwinproject.ac.uk/entry-2086

p. 61   "Wallace replied that he was glad..." – Darwin Correspondence Project. Letter 2145. http://www.darwinproject.ac.uk/entry-2145

p. 61   "Darwin was not prepared..." – Darwin Correspondence Project. Letter 2285. http://www.darwinproject.ac.uk/entry-2285

p. 62   "Darwin's confidants..." – The Complete Work of Charles Darwin Online. On the Tendency of Species to form Varieties; and on the Perpetuation of Varieties and Species by Natural Means of Selection. http://darwin-online.org .uk/content/frameset?viewtype=text&itemID=F350&pageseq=1

p. 64   "Darwin opened his book..." – The Complete Work of Charles Darwin Online. On the Origin of Species by Means of Natural Selection (First Edition). http://darwin-online.org.uk/content/frameset?itemID=F373&viewtype=text &pageseq=1

p. 66   "That morning he wrote..." – Darwin Correspondence Project. Letter 2546. http://www.darwinproject.ac.uk/entry-2546

p. 66   "Sedgwick, not holding back..." – Darwin Correspondence Project. Letter 2548. http://www.darwinproject.ac.uk/entry-2548

p. 66   "Darwin almost pleaded…" – Darwin Correspondence Project. Letter 2544. http://www.darwinproject.ac.uk/entry-2544

p. 66   "One of the first attacks…" – The Complete Work of Charles Darwin Online. Review of *Origin* & other works. http://darwin-online.org.uk/content/frameset?viewtype=text&itemID=A30&pageseq=1

p. 67   "The Oxford geologist John Phillips…" – Phillips, John. 1860. Life on Earth. London: Macmillan and Co.

p. 67   "Similar appraisals were made by paleontologists abroad…" – Rudwick, Martin. 1985. The Meaning of Fossils. New York: Science History Publications, USA. pp. 218-261.

p. 68   "In an 1861 letter…" Darwin Correspondence Project. Letter 3081. http://www.darwinproject.ac.uk/entry-3081

FROM FINS TO FINGERS

p. 69   "It was an ugly fish…" – Zimmer, Carl. 1998. At the Water's Edge. New York: The Free Press. pp. 9-19.

p. 70   "In 1841 Owen announced…" Owen, Richard. 1841. Description of Lepidosiren. *Transactions of the Linnean Society of London.* 18: 327-355.

p. 71   "Owen's reaction to this…" – Darwin Correspondence Project. Letter 2575. http://www.darwinproject.ac.uk/entry-2575

p. 72   "The anatomical pattern Owen saw…" Owen, Richard. 2007. On the Nature of Limbs. Chicago: University of Chicago Press.

p. 75   "Once again it was Owen…" Desmond, Adrian. 1982. Archetypes and Ancestors. Chicago: University of Chicago Press. pp. 65-72.

p. 76   "In *On the Origin of Species*…" – The Complete Work of Charles Darwin Online. On the Origin of Species by Means of Natural Selection (First Edition). http://darwinonline.org.uk/content/frameset?itemID=F373&viewtype=text&pageseq=1

p. 76   "In his 1922 study…" – Gregory, W.K. 1922. The Origin and Evolution of the Human Dentition. Baltimore: Williams and Wilkins Company. pp. 9.

p. 77   "In his 1917 textbook…" Lull, Richard Swann. 1917. Organic Evolution. New York: The Macmillan Company. pp. 477-490.

p. 77   "The only idea that Lull…" – Barrell, Joseph. 1916. The Influence of Silurian-Devonian Climates on the Rise of Air-Breathing Vertebrates. *Bulletin of the Geological Society of America.* XXVII: 387-436.

p. 78   "From the last decades…" Bowler, Peter. 1983. The Eclipse of Darwinism. Baltimore: Johns Hopkins University Press.

p. 79   "The first signs of fossil creatures…" – Zimmer, Carl. 1998. At the Water's Edge. New York: The Free Press. pp. 30-34.

p. 81   "It was not so much…" – Romer, Alfred Sherwood. 1956. The Early Evolution of Land Vertebrates. *Proceedings of the American Philosophical Society* 100: 157-167.

p. 81   Romer, Alfred Sherwood. 1958. Tetrapod Limbs and Early Tetrapod Life. *Evolution* 12: 365-369.

p. 81   "Clack's hunt for another fossil..." - Zimmer, Carl. 1998. At the Water's Edge. New York: The Free Press. pp. 51-56.

p. 82   "*Acanthostega* was not an aberration..." – Coates, M. I., and J. A. Clack. 1990. Polydactyly in the Earliest Known Tetrapod Limbs. *Nature* 347: 66-69.

p. 83   "The question was where..." – Bowler, Peter. 1996. Life's Splendid Drama. Chicago: University of Chicago Press. pp. 244-250.

p. 83   "This is precisely what..." – Shubin, N., and Pere Alberch. 1986. A Morphogenetic Approach to the Origin and Basic Organization of the Tetrapod Limb. *Evolutionary Biology* 20: 319-387.

p. 84   "The ancient genetic..." – Tabin, Clifford. 1992. Why We Have (Only) Five Fingers Per Hand: Hox Genes and the Evolution of Paired Limbs. *Development* 116: 289-296.

p. 84   Ruvinsky, Ilya, and Jeremy Gibson-Brown. 2000. Genetic and Developmental Bases of Serial Homology in Vertebrate Limb Evolution. *Development* 127: 5233-5244.

p. 84   "This pattern was supported..." – Carroll, Robert. 2001. The Origin and Early Radiation of Terrestrial Vertebrates. *Journal of Paleontology*. 75: 1202-1213.

p. 85   "The presence of hands and feet..." Ahlberg, Per Erik, Jennifer Clack, and Henning Blom. 2005. The Axial Skeleton of the Devonian Tetrapod Ichthyostega. *Nature* 437: 137-140.

p. 85   Clack, Jennifer. 2006. The emergence of early tetrapods. Palaeogeography, Paleoclimatology, *Palaeoecology* 232: 167-189.

p. 85   "That *Acanthostega* probably had..." – Gould, Stephen Jay. 1993. Full of Hot Air. Eight Little Piggies. New York: W.W. Norton. pp. 109-120.

p. 86   "Together both *Acanthostega*..." – Clack, Jennifer. 2002. Gaining Ground. Bloomington: Indiana University Press.

p. 86   Clack, Jennifer. 2009. The Fish-Tetrapod Transition: New Fossils and Interpretations. *Evolution: Education and Outreach* 2: 213-223.

p. 86   Coates, Michael, Marcello Ruta, and Matt Friedman. 2008. Ever Since Owen: Changing Perspectives on the Early Evolution of Tetrapods. *Annual Review of Ecology, Evolution, and Systematics* 39: 571-592.

p. 86   Long, John, and Malcolm Gordon. 2004. The Greatest Step in Vertebrate History: A Paleobiological Review of the Fish-Tetrapod Transition. *Physiological and Biochemical Zoology* 77: 700-719.

p. 87   "Though classified as a fish..." – Bosivert, Catherine. 2005. The Pelvic Fin and Girdle of Panderichthys and the Origin of Tetrapod Locomotion. *Nature* 438: 1145-1147.

p. 87   Bosivert, Catherine. 2009. The Humerus of Panderitchthys in Three Dimensions and Its Significance in the Context of the Fsh-tetrapod Transition. *Acta Zoologica* 90: 297-305.

p. 87   "Looking for early tetrapods..." Daeschler, Edward, Jennifer Clack, and Neil Shubin. Late Devonian Tetrapod Remains from Red Hill, Pennsylvania, USA: How Much Diversity? *Acta Zoologica* 90: 306-317.

p. 87   "The first step…" – Shubin, Neil. 2008. Your Inner Fish. New York: Pantheon Books.

p. 89   "Dubbed *Tiktaalik rosae*…" – Daeschler, Edward, Neil Shubin, and Farish Jenkins Jr. 2006. A Devonian Tetrapod-like Fish and the Evolution of the Tetrapod Body Plan. *Nature* 440: 757-763.

p. 89   Shubin, Neil, Edward Daeschler, and Farish Jenkins Jr. 2006. The Pectoral Fin of Tiktaalik Roseae and the Origin of the Tetrapod Limb. *Nature* 440: 764-771.

p. 89   "When more complete skull…" Ahlberg, Per, Jennifer Clack, Ervins Luksevics, Henning Blom, Ivars Zupins. 2008. Ventastega Curonica and the Origin of Tetrapod Morphology. *Nature* 453: 1199-1204.

FOOTPRINTS AND FEATHERS ON THE SANDS OF TIME

p. 91   "Humans have been finding…" – Mayor, Adrienne. 2000. The First Fossil Hunters. Princeton: Princeton University Press.

p. 91   "At the time Europeans…" Mayor, Adrienne. 2005. Fossil Legends of the First Americans. Princeton: Princeton University Press.

p. 93   "Hitchcock was deeply inspired…" – Marche, Jordan. 1991. Edward Hitchcock's Poem, The Sandstone Bird (1836). *Earth Sciences History* 1: 5-8

p. 94   "In his *Ichnology of New England*…" – Hitchcock, Edward. 1858. Ichnology of New England. Boston: William White.

p. 95   "Birds were so different…" Switek, Brian. (in press) Thomas Henry Huxley and the reptile-to-bird transition. Dinosaurs (and Other Extinct Saurians): A Historical Perspective. London: Geological Society of London.

p. 96   "Owen's description…" Weishampel, David, and Nadine White. 2003. The Dinosaur Papers 1676-1906. Washington: Smithsonian Books. pp. 276-288.

p. 96   "In an 1863 letter…" – Darwin Correspondence Project. Letter 3899. http://www.darwinproject.ac.uk/entry-3899

p. 96   "In the fourth edition…" The Complete Work of Charles Darwin Online. On the Origin of Species by Means of Natural Selection (Fourth edition) http://darwinonline .org.uk/content/frameset?itemID=F385&viewtype=text&pageseq=1

p. 96   "Huxley began his scientific career…" – Desmond, Adrian. 1994. Huxley: From Devil's Disciple to Evolution's High Priest. Reading: Addison-Wesley.

p. 97   "Huxley reiterated this point…" Switek, Brian. (in press) Thomas Henry Huxley and the reptile-to-bird transition. Dinosaurs (and Other Extinct Saurians): A Historical Perspective. London: Geological Society of London.

p. 99   "This New World relative…" – Cope, E.D. 1867. The Fossil Reptiles of New Jersey. *The American Naturalist* 1: 23-30.

p. 100 "Speculating upon…" – Huxley, T. H. 1868. On the Animals Which are Most Nearly Intermediate Between Birds and Reptiles. *Geological Magazine* v: 357-365.

p. 101 "Huxley thought it reasonable…" Huxley, T. H. 1869. Further Evidence of the Affinity Between the Dinosaurian Reptiles and Birds. *The Quarterly Journal of the Geological Society of London* xxvi: 12-31.

p. 102   "As Richard Owen stated..." Owen, Richard. 1864. A monograph of a Fossil Dinosaur (Osmosaurus armatus). *Paleontographical Society Monographs* 45-93.

p. 102   "To prevent this sort of confusion..." – Huxley, T. H. 1870. The Anniversary Address of the President. *The Quarterly Journal of the Geological Society of London* xxvi: xxix-lxiv.

p. 103   "Along with *Archaeopteryx*..." Huxley, T. H. 1877. American Addresses. London: Macmillan and Co.

p. 105   "Taking a dinosaurian ancestry..." Williston, S. W. 1879. Are birds derived from dinosaurs? The Kansas City Review of Science and Industry (Volume Three). Kansas City: Ramsey, Millet & Hudson. pp. 457-460.

p. 106   "Beebe introduced his colleagues..." – Beebe, William. 1915. A Tetrapteryx Stage in the Ancestry of Birds. *Zoologica* 2: 37-52.

p. 107   "The Danish artist..." Heilmann, Gerhard. 1926. The Origin of Birds. New York: D. Appleton and Company.

p. 108   "They called the new predator *Deinonychus*..." Ostrom, John. 1969. Osteology of Deinonychus antirrhopus, an unusual theropod from the Lower Cretaceous of Montana. *Bulletin of the Peabody Museum of Natural History* 30.

p. 109   "What was supposed about their biology..." Colbert, Edwin, Raymond Cowles, and Charles Bogert. 1946. Temperature tolerances in the American alligator and their bearing on the habits, evolution, and extinction of the dinosaurs. *Bulletin of the American Museum of Natural History* 86: 327-374.

p. 110   "After simmering for several years..." – Thomas, Roger D. K., and Everett Olson. A Cold Look at the Warm-Blooded Dinosaurs. Boulder: Westview Press.

p. 111   "After carefully studying..." – Ostrom, John. 1974. Archaeopteryx and the Origin of Flight. *The Quarterly Review of Biology* 49: 27-47.

p. 112   "The new dinosaur..." Ji, Q., and S. Ji. 1996. On discovery of the earliest bird fossil in China and the origin of birds. *Chinese Geology* 10: 30-33.

p. 112   Chen, P., Z. Dong, S. Zhen. 1998. An exceptionally well-preserved theropod dinosaur from the Yixian Formation of China. *Nature* 391: 147-152.

p. 112   "*Sinosauropteryx* was only the first..." – Chiappe, Luis M. 2007. Glorified Dinosaurs. Hoboken: John Wiley and Sons.

p. 112   "The fossil feathers of the strange..." – Schweitzer, M. H., J. A. Watt, R. Avci, L. Knapp, L. Chiappe, M. Norell, and M. Marshall. 1999. Beta-keratin specific immunological reactivity in feather-like structures of the Cretaceous alvarezsaurid, Shuvuuia deserti. *Journal of Experimental Zoology* 285: 146-157.

p. 112   Turner, A. H., P. J. Makovicky, and M .A. Norell. 2007. Feather quill knobs in the dinosaur Velociraptor. *Science* 317: 1721.

p. 113   "Both birds and alligators..." – Harris, M. P., J. F. Fallon, R. O. Prum. 2002. Shh-BMP2 signaling module and the evolutionary origin and diversification of feathers. *Journal of Experimental Zoology* 294: 160-176.

p. 113   "The diversity of feather types..." – Prum, Richard, and Alan Brush. 2002. The evolutionary origin and diversification of feathers. *The Quarterly Review of Biology* 77: 261-295.

p. 114   "One of the first tests…" – Vinther, Jakob, Derek Briggs, Julia Clarke, Gerald Mayr, and Richard Prum. 2009. Structural coloration in a fossil feather. *Biology Letters* 6: 128-131.

p. 114   "The first team…" – Zhang, Fucheng, Stuart Kearns, Patrick Orr, Michael Benton, Zhonghe Zhou, Diane Johnson, Zing Xu, and Xiaolin Wang. 2010. Fossilized melanosomes and the colour of Cretaceous dinosaurs and birds. *Nature* 463: 1075:1078.

p. 114   "Vinther and his team…" Li, Quanguo, Ke-Qin Gao, Jakob Vinther, Matthew Shawkey, Julia Clarke, Liliana D'Alba, Qingjin Meng, Derek Briggs, and Richard Prum. 2010. Plumage color patterns of an extinct dinosaur. *Science* 327: 1369-1372.

p. 115   "In 2002 Gerlad Mayr…" – Mayr, Gerald, D. Stefan Peter, Gerhard Plodowski, Olaf Vogel. 2002. Bristle-like integumentary structures at the tail of the horned dinosaur Psittacosaurus. *Naturwissenschaften* 89: 361-365.

p. 115   "It was joined in 2009…" – Zheng, Xiao-Ting, Hai-Lu You, Xing Xu, and Zhi-Ming Dong. 2009. An Early Cretaceous heterodontosaurid dinosaur with filamentous integumentary structures. *Nature* 458: 333-336.

p. 117   "As discovered by scientist Kenneth Dial…" Dial, Kenneth, Brandon Jackson, and Paolo Segre. 2008. A fundamental avian wing-stroke provides a new perspective on the evolution of flight. *Nature* 451: 985-990.

p. 117   "Described in 2002…" Czerkas, S.A., and C. Yuan. 2002. An arboreal maniraptoran from northeast China. Feathered Dinosaurs and the Origin of Flight. Blanding: The Dinosaur Museum. pp. 63-95.

p. 117   "Its description was followed…" – Zhang, Fucheng, Zhonghe Zhou, Xing Xu, Xiaolin Wang, and Corwin Sullivan. 2008. A bizarre Jurassic maniraptoran from China with elongate ribbon-like feathers. *Nature* 455: 1105-1108.

p. 118   "The unstable relationships…" Xu, Xing, Q. Zhao, M. Norell, C. Sullivan, D. Hone, G. Erickson, X. Wang, F. Han. 2009. A new feathered maniraptoran dinosaur fossil that fills a morphological gap in avian origin. *Chinese Science Bulletin* 54: 430-435.

p. 118   Hu, D., L. Hou, L. Zhang, X. Xu. 2009. A pre-Archaeopteryx troodontid theropod from China with long feathers on the metatarsus. *Nature* 461: 640-463.

p. 119   "But in 1994…" Norell, M.A., J.M. Clark, D. Dashzeveg, T. Barsbold, L.M. Chiappe, A.R. Davidson, M.C. McKenna, and M.J. Novacek. 1994. A theropod dinosaur embryo, and the affinities of the Flaming Cliffs Dinosaur eggs. *Science* 266: 779-782.

p. 119   "This discovery of fossilized behavior…" – Xu, X., and M. Norell. 2004. A new troodontid dinosaur from China with avian-like sleeping posture. *Nature* 431: 838-841.

p. 120   "The unique breathing system…" – O'Conner, Patrick, and Leon Claessens. 2006. Basic avian pulmonary design and flow-through ventilation in non-avian theropod dinosaurs. *Nature* 436: 253-256.

p. 120   Sereno, P.C., R.N. Martinez, J.A. Wilson, D.J. Varricchio, O.A. Alcober, and Hans Larsson. 2009. Evidence for Avian Intrathoracic Air Sacs in a New

Predatory Dinosaur from Argentina. *PLoS One* 3: e3303 http://www.plosone
.org/article/info%3Adoi%2F10.1371%2Fjournal.pone.0003303

p. 120 Wedel, M.J. 2009. Evidence for Bird-Like Air Sacs in Saurischian Dinosaurs.
Journal of Experimental Zoology 311A: 611-628

p. 121 "Some dinosaurs were even plagued..." Wolff, Ewan, Steven Salisbury, John
Horner, and David Varricchio. 2009. Common avian infection plagued the
tyrant dinosaurs. *PLoS One* 4: e7288 http://www.plosone.org/article/info
%3Adoi%2F10.1371%2Fjournal.pone.0007288

p. 123 "In his 1871 critique..." – Mivart, G.J. 1871. On the Genesis of Species. London: Macmillan and Co.

THE MEEK INHERIT THE EARTH

p. 126 "As he searched the dusty..." Desmond, Adrian. 1982. Archetypes and Ancestors. Chicago: University of Chicago Press.

p. 128 "Strangely, however..." Huxley, T.H. 1879. On the characters of the pelvis in
the Mammalia, and the conclusion respecting the origin of mammals which
may be based on them. *Proceedings of the Royal Society* xxviii: 395-405.

p. 128 "At that time..." – Ritvo, Harriet. 1997. The Platypus and the Mermaid.
Cambridge: Harvard University Press.

p. 129 "The American paleontologist..." Marsh, O.C. 1898. The Origin of Mammals. *The American Journal of Science* (Volume VI). New Haven: Tuttle,
Morehouse, and Taylor Press. pp. 407-409,

p. 131 "The great 'Bone Rush'..." Bowler, Peter. 1996. Life's Splendid Drama.
Chicago: University of Chicago Press. pp. 297-312.

p. 132 "Among this radiation..." – Kemp, T.S. 2005. The Origin and Evolution of
Mammals. New York: Oxford University Press.

p. 132 "The therapsids are distinguishable..." Kemp, T.S. 2006. The origin and early
radiation of the therapsid mammal-like reptiles: a palaeobiological hypothesis.
Journal of Evolutionary Biology 4: 1231-1247.

p. 132 Rubige, Bruce, and Christian Sidor. 2001. Evolutionary patterns among Permo-
Triassic Therapsids. *Annual Review of Ecological Systems* 32: 449-480.

p. 134 "The bones of the mammalian..." – Fritzsch, B., and K.W. Beisel. 2001. Evolution
and development of the vertebrate ear. *Brain Research Bulletin* 55: 711-721.

p. 134 Gould, Stephen Jay. 1993. "An Earful of Jaw." Eight Little Piggies. New York:
W.W. Norton pp. 95-108.

p. 134 Kemp, T.S. 2007. Acoustic transformer function of the postdentary bones and
quadrate of a nonmammalian cynodon. *Journal of Vertebrate Paleontology*
27: 431-441.

p. 134 Wang, Yuanqing, Yaoming Hu, Jin Meng, Chuankui Li. 2001. An ossified
Meckel's cartilage in two Cretaceous mammals and origin of the mammalian
middle ear. *Science*. 294: 357-361.

p. 134 Watson, D.M.S. 1953. The evolution of the mammalian ear. *Evolution* 7: 159-
177.

p. 136 "A recently-described early mammal..." – Luo, Zhe-Xi, Peiji Chen, Gang Li, and Meng Chen. 2007. A new eutriconodont mammal and evolutionary development in early mammals. *Nature* 446: 288-293.

p. 137 "We owe another..." – Angielczyk, Kenneth. 2009. Dimetrodon is not a dinosaur: Using tree thinking to understand the ancient relatives of mammals and their evolution. Evolution: Education and Outreach 2: 257-271.

p. 137 Crompton, A.W. 1963. The evolution of the mammalian jaw. *Evolution* 17: 431-439.

p. 137 DeMar, Robert, and Herbert Barghusen. 1972. Mechanics and the evolution of the synapsid jaw. *Evolution* 26: 622-637.

p. 137 Erwin, Douglas. 2006. Extinction: How life on earth nearly ended 250 million years ago. Princeton: Princeton University Press.

p. 137 Parrington, F.R., and T.S. Westoll. 1940. On the evolution of the mammalian palate. Philosophical *Transactions of the Royal Society of London, Series B, Biological Sciences* 230: 305-355.

p. 137 Sidor, Christian. 2003. Evolutionary trends and the origin of the mammalian lower jaw. *Paleobiology* 29: 605-640.

p. 138 "Extinction came quickly..." – Benton, Michael. 2005. When Life Nearly Died. London: Thames & Hudson.

p. 138 Erwin, Douglas. 1998. The end of the beginning: recoveries from mass extinctions. *Trends in Ecology and Evolution* 13: 344-349.

p. 138 Knoll, Andrew, Richard Bambach, Jonathan Payne, Sara Pruss, Woodward Fischer. 2007. Paleophysiology and end-Permian mass extinction. *Earth and Planetary Science Letters* 256: 295-313.

p. 138 Nicolas, Merrill, and Bruce Rubidge. 2009. Changes in Permo-Triassic terrestrial tetrapod ecological representation in the Beaufort Group (Karoo Supergroup) of South Africa. *Lethaia* 43: 45-59.

p. 138 Payne, Jonathan, and Seth Finnegan. 2007. The effect of geographic range on extinction risk during background and mass extinction. *Proceedings of the National Academy of Sciences* 104: 10506-10511.

p. 138 Retallack, Gregory, Roger Smith, and Peter Ward. 2003. Vertebrate extinction across Permian-Triassic boundary in Karoo Basin, South Africa. *GSA Bulletin* 115: 1133-1152.

p. 138 Retallack, Gregory, Christine Metzger, Tara Greaver, A. Hope Jahren, Roger Smith, Nathan Sheldon. 2006. Middle-Late Permian mass extinction on land. *GSA Bulletin* 118: 1398-1411.

p. 138 Sahney, Sarda, and Michael Benton. 2007. Recovery from the most profound mass extinction of all time. *Proceedings of the Royal Society B* 275: 759-765.

p. 138 Ward, Peter, Jennifer Both, Roger Buick, Michiel O. De Kock, Douglas Erwin, Geoffrey Garrison, Joseph Kirschvink, Roger Smith. 2005. Abrupt and gradual extinction among Late Permian land vertebrates in the Karoo Basin, South Africa. *Science* 307: 709-714.

p. 138 Weissflog, L., N.F. Elansky, K. Kott, F. Keppler, A. Pfennigsdorff, C.A. Lange, E. Putz, L.V. Lisitsyna. 2009. Late Permian changes in conditions of the atmosphere and environments caused by halogenated gases. *Doklady Earth Sciences* 425: 291-295.

p. 141   "This is not to say…" Luo, Zhe-Xi. 2007. Transformation and diversification in early mammal evolution. *Nature* 450: 1011-1019.

p. 143   "Marsupial and placental…" Ji, Qiang, Zhe-Xi Luo, Chong-Xi Yuan, John Wible, Jian-Ping Zhang, and Justin Georgi. 2002. The earliest known eutherian mammal. *Nature* 416: 816-822.

p. 144   "Then came another great extinction…" – Schulte, Peter, Laia Alegret, Ignacio Arenillas, José A. Arz, Penny J. Barton, Paul R. Bown, Timothy J. Bralower, Gail L. Christeson, Philippe Claeys, Charles S. Cockell, Gareth S. Collins, Alexander Deutsch, Tamara J. Goldin, Kazuhisa Goto, José M. Grajales-Nishimura, Richard A. F. Grieve, Sean P. S. Gulick, Kirk R. Johnson, Wolfgang Kiessling, Christian Koeberl, David A. Kring, Kenneth G. MacLeod, Takafumi Matsui, Jay Melosh, Alessandro Montanari, Joanna V. Morgan, Clive R. Neal, Douglas J. Nichols, Richard D. Norris, Elisabetta Pierazzo, Greg Ravizza, Mario Rebolledo-Vieyra, Wolf Uwe Reimold, Eric Robin, Tobias Salge, Robert P. Speijer, Arthur R. Sweet, Jaime Urrutia-Fucugauchi, Vivi Vajda, Michael T. Whalen, and Pi S. Willumsen. 2010. The Chicxulub asteroid impact and mass extinction at the Cretaceous-Paleogene Boundary. *Science* 327: 1214-1218.

As Monstrous as a Whale

p. 145   "In the summer of 1841…" Semonin, Paul. 2000. American Monster. New York: New York University Press.

p. 146   "By 1844, however…" – Koch, Albert. 1972. Journey through a Part of the United States of North America in the Years 1844-1846. Carbondale: Southern Illinois University Press.

p. 150   "The *New York Dissector*…" Anonymous. 1848. New York Dissector (Volume 1). New York: Henry Hall Sherwood pp. 217-223.

p. 151   "The Harvard anatomist…" Buckley, S.B. 1846. On Zeuglodon Remains of Alabama. *The American Journal of Science and Arts* (Vol II). New Haven: B. L. Hamlen pp. 125-133.

p. 152   "While entertaining some naturalists…" Anonymous. 1862. Foreign Intelligence. The Geologist (Volume 5) London: Lovell Reeve & Co. pp. 105.

p. 152   "Curiously, the controversy…" Thomson, Keith. 2008. The Legacy of the Mastodon. New Haven: Yale University Press. pp. 83-85.

p. 153   "Comparative anatomy…" Harlan, Richard. 1835. Medical and Physical Researches. Philadelphia: Lydia Bailey. pp. 282-283.

p. 155   "Based upon observations…" - The Complete Work of Charles Darwin Online. On the Origin of Species by Means of Natural Selection (First Edition). http://darwin-online.org.uk/content/frameset?itemID=F373&viewtype=text&pageseq=1

p. 156   "As related to Charles Lyell…" – Darwin Correspondence Project. Letter 2575. http://www.darwinproject.ac.uk/entry-2575

p. 156   "In an 1883 lecture…" – Flower, W. H. 1898. Essays on Museums. London: Macmillan. pp. 209-231.

p. 157   "As E.D. Cope admitted…" Cope, E.D. 1890. The Cetacea. *The American Naturalist* XXIV: 599-616.

p. 158   "While revising the relationships…" Van Valen, Leigh. 1966. Deltatheridia, a new order of mammals. *American Museum of Natural History Bulletin* 132:1-126.

p. 159   "This lack of early transitional…" Lipps, Jere, and Edward Mitchell. 1976. Model for the adaptive radiations and extinctions of pelagic marine mammals. *Paleobiology* 2: 147-155.

p. 159   "A startling discovery…" Gingerich, P. D., D. E. Russell. 1981. Pakicetus inachus, a new archaeocete (Mammalia, Cetacea) from the early-middle Eocene Kuldana Formation of Kohat (Pakistan). Contributions from the Museum of Paleontology: The University of Michigan. 25: 235-246.

p. 160   "There were still things to learn…" Gingerich, Philip, B. Holly Smith, and Elwyn Simons. 1990. Hind Limbs of Eocene Basilosaurus: Evidence of feet in whales. *Science* 249: 154-157.

p. 160   "The skeleton of *Pakicetus attocki*…" Thewissen, J.G.M, E.M. Williams, L.J. Rose, and S.T. Hussain. 2001. Skeletons of terrestrial cetaceans and that relationship of whales to artiodactyls. *Nature* 413: 277-281.

p. 161   "The earliest known archaeocetes…" Thewissen, J. G. M. 1998. The Emergence of Whales. New York: Plenum Press.

p. 163   "A better candidate…" Thewissen, J. G. M. 2007. Whales originated from aquatic artiodactyls in the Eocene epoch of India. *Nature* 450: 1190-1194.

p. 165   "When the fossil data…" Geisler, Jonathan, and Jessica Theodor. 2009. Hippopotamus and whale phylogeny. Nature 458: E1-E4.

p. 165   "There were many types of marine reptiles…" Motani, Ryosuke. 2009. The evolution of marine reptiles. *Evolution: Education and Outreach* 2: 224-235.

p. 165   "One particular group…" Motani, Ryosuke. 2005. Evolution of fish-shaped reptiles (Reptilia: Ichthyopterygia) in their physical environments and constraints. *Annual Review of Earth and Planetary Sciences* 33: 395-420.

p. 167   "This vertebral adaptation…" Gingerich, Philip, S. Mahmood Raza, Muhammad Arif, Mohammad Anwar, and Xiaoyuan Zhou. 1994. New whale from the Eocene of Pakistan and the origin of cetacean swimming. *Nature* 368: 844-847.

p. 167   Gingerich, P.D., Munir ul-Haq, Wighart von Koenigswalk, William Sanders, B. Holly Smith, and Iyad Zalmout. 2009. New protocetid whales from the Middle Eocene of Pakistan: Birth on land, precocial development, and sexual dimorphism. PLoS One 4: e4366 http://www.plosone.org/article/info%3Adoi %2F10.1371%2Fjournal.pone.0004366

p. 168   "*Georgiacetus* had relatively…" Uhen, Mark. 2008. New protocetid whales from Alabama and Mississippi, and a new cetacean clade, Pelagiceti. *Journal of Vertebrate Paleontology* 28: 589-593.

p. 170   "At twenty-four days old…" Thewissen, J .G. M., M. J. Cohn, L. S. Stevens, S. Bajpai, J. Heyning, and W.E. Horton Jr. 2006. Developmental basis for hind-limb loss in dolphins and origin of the cetacean bodyplan. *Proceedings of the National Academy of Sciences* 103: 8414-8418.

p. 171   "In the case of living mysticetes..." Demere, Thomas, Michael McGowen, Annalisa Berta, and John Gatesy. 2008. Morphological and molecular evidence for a stepwise evolutionary transition from teeth to baleen in mysticete whales. *Systematic Biology* 57: 15-37.

BEHEMOTH

p. 175   "Buffon had fired..." – Semonin, Paul. 2000. American Monster. New York: New York University Press.

p. 177   "The graveyard..." – Mayor, Adrienne. 2005. Fossil Legends of the First Americans. Princeton: Princeton University Press.

p. 178   "If Lewis and Clark..." – Hedeen, Stanley. 2008. Big Bone Lick. Lexington: The University Press of Kentucky.

p. 179   "In 1796 Cuvier..." – Rudwick, Martin. 1997. Georges Cuvier. Chicago: University of Chicago Press.

p. 181   "Even more fossil elephants..." Cohen, Claudine. 1994. The Fate of the Mammoth. Chicago: University of Chicago Press.

p. 184   "At the turn of the twentieth century..." – Simons, Elwyn. 2008. Eocene and Oligocene mammals of the Fayum, Egypt. Elwyn Simons: A Search for Origins. New York: Springer. pp. 87-105.

p. 185   "Thus the great march..." Barbour, Erwin. 1914. Mammalian fossils from Devil's Gulch. University Studies. Lincoln: University of Nebraska. pp. 185-202.

p. 187   "The task of reinterpreting..." Simpson, G.G. 1945. On the Principles of Clas sification and the Classification of Mammals. *Bulletin of the American Museum of Natural History* 85.

p. 188   "Manatees and elephants..." – Domning, Daryl, Clayton Ray, and Malcolm McKenna. 1986. Two new Oligocene desmostylians and a discussion of tethytherians systematic. *Smithsonian Contributions to Paleobiology* 59: 1-56.

p. 189   "The first animals..." Gheerbrant, Emmanual. 2009. Paleocene emergence of elephant relatives and the rapid radiation of African ungulates. *Proceedings of the National Academy of Sciences* 106: 10717-10721.

p. 190   "In 2007..." Liu, Alexander, Erik Seiffert, and Elwyn Simons. 2008. Stable isotope evidence for an amphibious phase in early proboscidean evolution. *Proceedings of the National Academy of Sciences* 105: 5786-5791.

p. 190   "This masticatory transition..." Shoshani, Jeheskel, Robert C. Walter, Michael Abraha, Seife Berhe, Pascal Tassy, William J. Sanders, Gary H. Marchant, Yosief Libsekal, Tesfalidet Ghirmai, and Dietmar Zinner. A proboscidean from the late Oligocene of Eritrea, a "missing link" between early Elephantiformes and Elephantimorpha, and biogeographic implications. *Proceedings of the National Academy of Sciences* 103: 17296-17301.

p. 192   "Proboscideans remained..." – Shoshani, Jeheskel. 1998. Understanding proboscidean evolution: a formidable task. *Trends in Ecology and Evolution* 13: 480-487.

p. 192   Shoshani, Jeheskel, and Pascal Tassy. 2005. Advances in proboscidean taxonomy & classification, anatomy & physiology, and ecology & behavior. *Quaternary International* 126-128: 5-20.

p. 193   "Though the genus *Mammuthus...*" – Agenbroad, Larry. 2005. North American Proboscideans: Mammoths: The state of Knowledge, 2003. *Quaternary International* 126-128: 73-92.

p. 193   Lister, Adrian, Andrei Sher, Hans van Essen, Guangbiao Wei. 2005. The pattern and process of mammoth evolution in Eurasia. *Quaternary International* 126-128: 49-64.

p. 194   "Punctuated equilibrium..." Eldredge, Niles, and Stephen Jay Gould. 1972. Punctuated equilibria: An alternative to phyletic gradualism. Models in Paleobiology. San Francisco: Freeman, Cooper and Company. pp 82-115.

p. 195   "The continuously shifting..." Alvarez-Lao, Diego, Ralf-Dietrich Kahlke, Nuria Garcia, and Dick Mol. 2009. The Padul mammoth finds – On the southernmost record of Mammuthus primigenius in Europe and its southern spread during the Late Pleistocene. *Palaeogeography, Palaeoclimatology, Palaeoecology* 278: 57-70.

p. 195   "Around 47,000 years..." Agenbroad, L.D. 2001. Channel Islands (USA) pygmy mammoths (Mammuthus exilis) compared and constrasted with M. columbi, their continental ancestral stock. The World of Elephants: Proceedings of the 1st International Congress Rome: Consiglio Nazionale delle Ricerche. pp. 473-475.

p. 195   Agenbroad, L.D. 2002. New localities, chronology, and comparisons for the pygmy mammoth (Mammuthus exilis): 1994-1998. Proceedings of the Fifth California Islands Symposium. Santa Barbara: Santa Barbara Museum of Natural History. pp. 518-524.

p. 196   "Other populations..." – Guthrie, R. Dale. 2004. Radiocarbon evidence of mid-Holocene mammoths stranded on an Alaskan Bering Sea island. *Nature* 429: 746-749.

p. 196   Roth, V.I. 2001. Ecology and evolution of dwarfing in insular elephants. The World of Elephants: Proceedings of the 1st International Congress Rome: Consiglio Nazionale delle Ricerche. pp. 507-509.

p. 196   Stuart, Anthony, Leopold Sulerzhitsky, Lyobov Orlova, Yaroslav Kuzmin, and Adrian Lister. 2002. The latest woolly mammoths (Mammuthus primigenius Blumenbach) in Europe and Asia: a review of the current evidence. *Quaternary Science Reviews* 21: 1559-1569.

p. 196   "As mammoths spread..." Arroyo-Cabrales, Joaquin, Oscar Polaco, Cesar Laurito, Eileen Johnson, Maria Teresa Alberdi, Ana Lucia Valerio Zamora. 2007. The proboscideans (Mammalia) from Mesoamerica. *Quaternary International* 169-170: 17-23.

p. 196   Baskin, J.A., and R.G. Thomas. 2007. South Texas and the Great American Interchange. *Gulf Coast Association of Geological Societies Transactions* 57: 37-45.

p. 196   MacFadden, Bruce. 2006. Extinct mammalian biodiversity of the ancient New World tropics. *Trends in Ecology and Evolution* 21: 157-165.

p. 196 Webb, S. David. 1991. Ecogeography and the Great American Interchange. *Paleobiology* 17: 266-280

p. 198 "Numerous non-mutually exclusive..." – Cione, Alberto, Eduardo Tonni, and Leopoldo Soibelzon. 2009. Did humans cause the Late Pleistocene-Early Holocene mammalian extinctions in South America in a context of shrinking open areas? American Megafaunal Extinctions at the End of the Pleistocene. New York: Springer. pp. 125-144

p. 198 Fisher, Daniel. 2009. Paleobiology and Extinction of Proboscideans in the Great Lakes region of North America. American Megafaunal Extinctions at the End of the Pleistocene. New York: Springer. pp. 55-75

p. 198 Gaudzinskia, S., F. Turnera, A.P. Anzideib, E. Alvarez-Fernandezc, J. Arroyo-Cabralesd, J. Cinq-Marse, V.T. Dobosif, A. Hannusg, E. Johnsonh, S.C. M.unzeli, A. Scheeri, P. Villaj. 2005. The use of Proboscidean remains in every-day Palaeolithic life. *Quaternary International* 126-128: 179-194.

p. 198 Grayson, Donald. 2008. Holocene underkill. *Proceedings of the National Academy of Sciences* 105: 4077-4078.

p. 198 Haynes, Gary. 2007. A review of some attacks on the overkill hypothesis, with special attention to misrepresentations and doubletalk. *Quaternary International* 169-170: 84-94.

p. 198 Koch, Paul, and Anthony Barnosky. 2006. Late Quaternary Extinctions: State of the Debate. *Annual Reviews of Ecological and Evolutionary Systems.* 37: 215-250.

p. 198 Kuzmin, Ya., I..A. Orlova, and V.N. Dementiev. 2008. Dynamics of Mammoth (Mammuthus primigenius Blum.) populations of Asia and North America and its correlation with climatic changes in the Late Neopleistocene (45 000-9700 Years BP) *Doklady Earth Sciences* 421A: 978-982.

p. 198 Haile, J., D. Froese, R. MacPhee, R. Roberts, L. Arnold, A. Reyes, M. Rasmussen, R. Nielsen, B. Brook, S. Robinson, M. Demuro, M. Gilbert, K. Munch, J. Austin, A. Cooper, I. Barnes, P. Möller, and E. Willerslev. 2009. Ancient DNA reveals late survival of mammoth and horse in interior Alaska. *Proceedings of the National Academy of Sciences* 106: 22352-22357.

p. 198 Lister, Adrian, and Anthony Stuart. 2008. The impact of climate change on large mammal distribution and extinction: Evidence from the last glacial/interglacial transition. *C.R. Geoscience* 340: 615-620.

p. 198 Morrison, John, Wes Sechrest, Eric Dinerstein, David Wilcove, and John Lamoreux. 2007. Persistence of large mammal faunas as indicators of global human impacts. *Journal of Mammalogy* 88: 1363-1380.

p. 198 Mussi, Margherita, and Paola Villa. 2008. Single carcass of Mammuthus primigenius with lithic artifacts in the Upper Pleistocene of northern Italy. *Journal of Archaeological Science* 35: 2606-2613.

p. 198 Nogues-Bravo, David, Jesus Rodriguez, Joaquin Hortal, Persaram Batra, Miguel Araujo. 2008. Climate change, humans, and the extinction of the woolly mammoth. *PLoS Biology* 6: e79 http://www.plosbiology.org/article/info:doi/10.1371/journal.pbio.0060079

p. 198   Putshkov, P.V. 2001. "Proboscidean agent" of some Tertiary megafaunal extinctions. The World of Elephants: Proceedings of the 1st International Congress Rome: Consiglio Nazionale delle Ricerche. pp. 133-136.

p. 198   Stuart, Anthony. 2005. The extinction of woolly mammoth (Mammuthus primigenius) and straight-tusked elephant (Palaeoloxodon antiques) in Europe. *Quaternary International* 126-128: 171-177.

p. 198   Stuart, A.J., Kosintsev, T.F.G Higham, and A.M. Lister. 2004. Pleistocene to Holocene extinction dynamics in giant deer and woolly mammoth. *Nature* 431: 684-689.

p. 201   "But what if..." Caro, Tim. 2007. The Pleistocene re-wilding gambit. *Trends in Ecology and Evolution* 22: 281-283.

p. 201   Donlan, Josh, Harry Greene, Joel Berger, Carl Bock, Jane Bock, David Burney, James Estes, Dave Foreman, Paul Martin, Gary Roemer, Felisa Smith, Michael Soule. 2005. Re-wilding North America. *Nature* 436: 913-914.

p. 201   Marris, Emma. 2009. Reflecting the past. *Nature* 462: 30-32.

p. 2010 "After all..." Orlando, Ludovic, Catherine Hanni, and Cristophe Douday. 2007. Mammoth and elephant phylogenetic relationships: Mammut americanum, the missing outgroup. *Evolutionary Bioinformatics* 2: 45-51.

p. 201   Roca, Alfred. 2008. The mastodon mitochondrial genome: a mammoth accomplishment. *Trends in Genetics* 24: 49-52.

p. 201   Rohland, Nadin, Anna-Sapfo Malaspinas, Joshua Pollack, Montgomery Slatkin, Paul Matheus, and Michael Hofreiter. 2007. Proboscidean mitogenomics: Chronology and mode of elephant evolution using mastodon as outgroup. *PLoS Biology* 5: e207 http://www.plosbiology.org/article/info:doi/ 10.1371/journal.pbio.0050207

ON A LAST LEG

p. 205   "In Gosse's next book..." Gosse, P.H. 1857. Omphalos. London: R. Clay.

p. 206   "Despite the massive amount..." Buffetaut, Eric. A Short History of Vertebrate Palaeontology. London: Croom Helm. 114-121.

p. 207   "It would not be long..." MacFadden, Bruce. 1992. Fossil Horses: Systematics, Paleobiology, and Evolution of the Family Equidae. New York: Cambridge University Press.

p. 208   "Kovalevskii's general scrics..." – Huxley, T.H. 1870. The Anniversary Address of the President. *The Quarterly Journal of the Geological Society of London* xxvi: xxix-lxiv.

p. 208   "The fossil horses..." – Huxley, T.H. 1877. American Addresses. London: Macmillan and Co.

p. 209   "Marsh's professional career..." Wallace, David Rains. 2004. Beasts of Eden. Berkeley: University of California Press.

p. 214   "In a 1907 review..." Gidley, James Williams. 1907. Revision of the Miocene and Pliocene Equidae of North America. *Bulletin of the American Museum of Natural History* XXIII: pp. 865-934.

p. 214   "Walter Granger..." – Granger, Walter. 1908. A revision of the American Eocene horses. *Bulletin of the American Museum of Natural History.* XXIV: 221-264.

p. 214   "In 1924 W. D. Matthew..." – Matthew, W. D. 1924. A new link in the ancestry of the horse. *American Museum Novitates* 131: 1-2.

p. 215   "The tension between..." – Matthew, W. D. The evolution of the horse: A record and its interpretation. The Quarterly Review of Biology 1 :139-185.

p. 216   "Matthew's emphasis..." Simpson, G.G. 1951. Horses. New York: Oxford University Press.

p. 216   "One such litoptern..." – Scott, W. B. 1913. A History of Land Mammals in the Western Hemisphere. New York: The Macmillian Company. pp 501-508.

p. 218   "The disarray..." – Froehlich, David. 2002. Quo vadis eohippus? The systematic and taxonomy of the early Eocene equids (Perissodactyla). *Zoological Journal of the Linnean Society* 134: 141-256.

p. 218   "In 1981 paleontologist..." – Gingerich, Philip. 1981. Variation, sexual dimorphism, and social structure in the Early Eocene horse Hyracotherium (Mammalia, Perissodactyla). *Paleobiology* 7: 443-455.

p. 218   "*Eohippus* and its close relatives..." – MacFadden, Bruce. 2005. Fossil horses – Evidence for evolution. *Science* 307: 1728-1730.

p. 220   "At first this..." – Prothero, Donald, and Neil Shubin. The Evolution of Oligocene Horses. The Evolution of Perissodactyls. New York: Oxford University Press. pp. 142-175.

p. 221   "It was a time..." – Hulbert Jr., Richard. 1993. Taxonomic evolution in North American Neogene horses (Subfamily Equinae): The rise and fall of an adaptive radiation. *Paleobiology* 19: 216-234.

p. 221   MacFadden, Bruce. 1985. Patterns of phylogeny and rates of evolution in fossil horses: Hipparions from the Miocene and Pliocene of North America. *Paleobiology* 11: 245-257.

p. 221   "Chewing grass..." – Bobe, Rene, and Anna Behrensmeyer. 2004. The expansion of grassland ecosystems in Africa in relation to mammalian evolution and the origin of the genus Homo. *Palaeogeography, Palaeoclimatology, Palaeoecology* 207: 399-420.

p. 221   MacFadden, Bruce. 2000. Cenozoic mammalian herbivores from the Americas: Reconstructing ancient diets and terrestrial communities. *Annual Reviews of Ecological Systematics* 31: 33-59.

p. 221   Radinsky, Leonard. 1984. Ontogeny and phylogeny in horse skull evolution. Evolution 38: 1-15.

p. 221   Stromberg, Caroline. 2006. Evolution of hypsodonty in equids: testing a hypothesis of adaptation. *Paleobiology* 32: 236-258.

p. 222   "A look at the evolution..." Gould, Stephen Jay. 1991. Life's Little Joke. Bully for Brontosaurus. New York: W.W. Norton. pp. 168-181.

p. 222   Gould, Gina., and Bruce MacFadden. 2004. Gigantism, Dwarfism, and Cope's Rule: "Nothing in evolution makes sense without a phylogeny." *Bulletin of the American Museum of Natural History* 285: 219-237.

p. 222   MacFadden, Bruce. 1986. Fossil horses from "Eohippus" (Hyracotherium) to Equus: Scaling, Cope's Law, and the Evolution of Body Size. *Paleobiology* 12: 355-369.

p. 223   "For hundreds of years..." Marsh, O. C. 1879. Polydactyle horses, recent and extinct. *American Journal of Science* 17: 499-505.

p. 223   Marsh, O.C. 1892. Recent polydactyle horses. *American Journal of Science* XLIII: 339-355.

THROUGH THE LOOKING GLASS

p. 226   "In 1655..." – Livingstone, David. 2008. Adam's Ancestors. Baltimore: The Johns Hopkins University Press. pp. 26-51.

p. 227   "In 1698..." – Corbey, Raymond. 2005. The Metaphysics of Apes. New York: Cambridge University Press. pp. 36-59.

p. 230   "In 1857..." – Owen, Richard. 1858. On the Characters, Principles of Division and Primary Groups of the Class Mammalia. *Proceedings of the Linnean Society of London* II: 1-37.

p. 230   "Charles Darwin balked..." – Darwin Correspondence Project. Letter 2117. http://www.darwinproject.ac.uk/entry-2117

p. 230   "Owen remained steadfast..." – Rupke, Nicolaas. 2009. Richard Owen: Biology Without Darwin. Chicago: University of Chicago Press.

p. 231   "Just how old..." – North, F. J. 1942. Paviland cave, the "Red Lady", the Deluge, and William Buckland. *Annals of Science* 5: 91-128.

p. 23   Sommer, Marianna. 2004. 'An amusing account of a cave in Wales': William Buckland (1784-1856) and the Red Lady of Paviland. *The British Journal for the History of Science* 37: 53-74.

p. 231   Van Riper, A. Bowdoin. 1993. Men Among the Mammoths. Chicago: University of Chicago Press.

p. 233   "This point..." – Owen, Richard. 1846. A History of British Fossil Mammals, and Birds. London: John Van Voorst. pp. 182-183.

p. 233   "The complexities..." Van Riper, A. Bowdoin. 1993. Men Among the Mammoths. Chicago: University of Chicago Press.

p. 237   "Even T.H. Huxley..." – Huxley, T. H. 1863. Evidence as to Man's Place in Nature. New York: D. Appleton and Company. pp. 181-182.

p. 237   "He certainly could have used..." Complete Work of Charles Darwin Online. Descent of Man and Selection in Relation to Sex (First edition). http://darwin-online.org.uk/content/frameset?itemID=F937.1&viewtype=text&pageseq=1

p. 238   "The young Dutch..." – Shipman, Pat. 2001. The Man Who Found the Missing Link. New York: Simon & Schuster.

p. 239   "This led some..." – Dubois, Eugene. 1898. Pithecanthropus erectus – a form from the ancestral stock of mankind. Adam or Ape: a Sourcebook. Cambridge: Schenkman Publishing Company. pp. 165-175.

p. 240   "Using the femur..." Gould, Stephen Jay. 1993. Men of the Thirty-Third Division: An Essay on Integrity. Eight Little Piggies. New York: W.W. Norton. pp. 124-137.

p. 240 "In 1908..." Spencer, Frank. 1990. Piltdown: A Scientific Forgery. New York: Oxford University Press.

p. 241 "In 1921..." Black, Davison. 1921. Tertiary Man in Asia: the Chou Kou Tien Discovery. Adam or Ape: a Sourcebook. Cambridge: Schenkman Publishing Company. pp. 219-220.

p. 241 Schmalzer, Sigrid. 2009. The People's Peking Man. Chicago: University of Chicago Press.

p. 242 "While teaching anatomy..." – Dart, Raymond, and Dennis Craig. 1959. Adventures with the Missing Link. New York: The Viking Press.

p. 243 "Dart rushed..." Dart, Raymond. 1925. Australopithecus africanus: The man-ape of South Africa. Adam or Ape: a Sourcebook. Cambridge: Schenkman Publishing Company. pp. 201-209.

p. 244 "Armed with copious notes..." – Le Gros Clark, W. E. 1947. Observations on the anatomy of the fossil Australopithecinae. Adam or Ape: a Sourcebook. Cambridge: Schenkman Publishing Company. pp. 293-314.

p. 244 Le Gros Clark, W.E. 1967. Man-apes or Ape-men? New York: Holt, Rinehart, and Winston.

p. 245 "Le Gros Clark's conclusions..." Gundling, Tom. 2005. First in Line: Tracing Our Ape Ancestry. New Haven: Yale University Press. pp. 127-140.

p. 245 Simpson, G.G. 1950. Some principles of historical biology bearing on human origins. Cold Spring Harbor Symposia on Quantitative Biology XV: 55-66.

p. 246 "Even more horrific..." Dart, Raymond. 1953. The Predatory Transition from Ape to Man. International Anthropological and Linguistic Review 1: 201-219.

p. 246 "In the summer of 1959 ..." Leakey, L. S. B. 1959. A new fossil skull from Olduvai. Adam or Ape: a Sourcebook. Cambridge: Schenkman Publishing Company. pp. 353-359.

p. 247 "This identification..." Leakey, L. S. B. 1961. Africa's Contribution to the Evolution of Man. The South African Archaeological Bulletin 16: 3-7.

p. 247 "If the job..." – Leakey, L. S. B., P. V. Tobias, and J. R. Napier. 1964. A new species of the genus Homo from Olduvai Gorge. Adam or Ape: a Sourcebook. Cambridge: Schenkman Publishing Company. pp. 431-438.

p. 248 "As Sherwood Washburn..." Lee, Richard, and Ireven Devore. 1968. Man the Hunter. Hawthorne: Aldine de Gruyter.

p. 250 "In 1967 ..." Sarich, V. M., and A. C. Wilson. 1967. Immunological time scale for hominid evolution. Science 158: 1200-1203.

p. 250 "Early work..." Johanson, Donald, and Maitland Edey. 1982. Lucy: The Beginnings of Humankind. New York: Warner Books.

p. 251 "They decided..." Johanson, D., T. D. White, and Y. Coppens. 1978. A new species of Australopithecus (Primates: Hominidae) from the Pliocene of East Africa. Kirtlandia 28.

p. 253 "At about the same..." Sibley, C. G., and J. E. Ahlquist. 1984. The phylogeny of the hominoid primates, as indicated by DNA-DNA hybridization. Journal of Molecular Evolution 20: 2-15.

p. 253   "Further studies..." Ijdo, J. W., A. Baldin, A. Baldin, D. C. Ward, S. T. Reeders, and R. A. Wells. 1991. Origin of human chromosome 2: An ancestral telomere-telomere fusion. *Proceedings of the National Academy of Sciences* 88: 9051-9055.

p. 254   "Even Dart's horrifying..." Brain, C. K. 1981. The Hunters or the Hunted? Chicago: University of Chicago Press.

p. 255   "Kimeu was part..." Walker, Alan, and Richard Leakey. 1993. The Nariokotome Homo erectus Skeleton. Netherlands: Springer.

p. 255   "What was needed..." – Simpson, Scott, Jay Quade, Naomi Levin, Robert Butler, Guillaume Dupont-Nivet, Melanie Everett, Sileshi Semaw. A female Homo erectus pelvis from Gona, Ethiopia. *Science* 322: 1089-1092.

p. 256   "In 2001..." Gibbons, Ann. 2006. The First Human. New York: Doubleday.

p. 256   Richmond, B.G. and W.L. Jungers. 2008. Orrorin tugenensis femoral morphology and the evolution of hominin bipedalism. *Science* 319: 1662-1665.

p. 256   Senut, B., Martin Pickford, Dominique Gommery, Pierre Mein, Kiptalam Cheboi, and Yves Coppens. 2001. First hominid from the Miocene (Lukeino Formation, Kenya). *Comptes Rendus de l'Academie de Sciences* 332: 137-144.

p. 256   "Another, even older..." Brunet, Michel, Franck Guy, David Pilbeam, Hassane Taisso Mackaye, Andossa Likius, Djimdoumalbaye Ahounta, Alain Beauvilain, Ce´ cile Blondel, Herve´ Bocherensk, Jean-Renaud Boisserie, Louis De Bonis, Yves Coppens, Jean Dejax, Christiane Denys, Philippe Duringerq, Vera Eisenmann, Gongdibe´ Fanone, Pierre Fronty, Denis Geraads, Thomas Lehmann, Fabrice Lihoreau, Antoine Louchart, Adoum Mahamat, Gildas Merceron, Guy Mouchelin, Olga Otero, Pablo Pelaez Campomanes, Marcia Ponce De Leon, Jean-Claude Rage, Michel Sapanet, Mathieu Schuster, Jean Sudre, Pascal Tassy, Xavier Valentin, Patrick Vignaud, Laurent Viriot, Antoine Zazzo, and Christoph Zollikofer. 2002. A new hominid from the Upper Miocene of Chad, Central Africa. *Nature* 418: 145-151.

p. 257   "Even though..." Moya-Sola, Salvador, Meike Kohler, David Alba, Isaac Casanovas-Vilar, and Jordi Galindo. 2004. Pierolapithecus catalaunicus, a new Middle Miocene great ape from Spain. *Science* 306: 1339-1344.

p. 257   Moya-Sola, S. D. Alba, S. Almecija, I. Casanovas-Vilar, M. Kohler, S. de Esteban-Trivigno, J. Robles, J. Galindo, and J. Fortuny. 2009. A unique Middle Miocene European hominoid and the origins of the great ape and human clade. *Proceedings of the National Academy of Sciences* 106: 9601-9606.

p. 257   Zollikofer, Cristoph, Marcia Ponce de Leon, Daniel Lieberman, Franck Guy, David Pilbeam, Andossa Likius, Hassane Mackaye, Patrick Vignaud, and Michel Brunet. 2005. Virtual cranial reconstruction of Sahelanthropus tchadensis. *Nature* 434: 755-759.

p. 258   "But there was..." Ashfaw, B., T. D. White, and G. Suwa. 1994. Australopithecus ramidus a new species of early hominid from Aramis, Ethiopia. *Nature* 371: 306.

p. 258   Haile-Selassie, Yohannes, Gen Suwa, and Tim White. 2004. Late Miocene teeth from Middle Awash, Ethiopia, and early hominid dental evolution. *Science* 303: 1503-1505.

p. 258 White, T. D., Berhane Asfaw, Yonas Beyene, Yohannes Haile-Selassie, C. Owen Lovejoy, Gen Suwa, and Giday WoldeGabriel. 2009. Ardipithecus ramidus and the paleobiology of early hominids. *Science* 326: 75-86.

p. 262 "In 2007..." Spoor, F., M.G. Leakey, P.N. Gathogo, F.H. Brown, S.C. Anton, I. McDougall, C. Kiarie, F.K. Manthi, and L.N. Leakey. 2007. Implications of new early Homo fossils from Ileret, east of Lake Turkana, Kenya. *Nature* 448: 688-691.

TIME AND CHANCE

p. 267 "An extended tenure..." Russell, D. A., and R. Seguin. 1982. Reconstructions of the small Cretaceous theropod Stenonychosaurus inequalis and a hypothetical dinosauroid. *Syllogeus* 37: 1-43.

p. 267 "We could feel...":

p. 267 Collins, Francis. 2006. The Language of God. New York: The Free Press.

p. 267 Conway Morris, Simon. 2003. Life's Solution. New York: Cambridge University Press.

p. 267 Miller, Kenneth. 2008. Only a Theory. New York: Viking.

p. 267 Russell, Dale. 2009. Islands in the Cosmos. Bloomington: Indiana University Press.

p. 268 "If we could somehow..." Gould, Stephen Jay. 1989. Wonderful Life. New York: W.W. Norton.

p. 269 "In 1988...," Blount, Zachary, Christina Borland, and Richard Lenski. 2008. Historical contingency and the evolution of a key innovation in an experimental population of Escherichia coli. *Proceedings of the National Academy of Sciences* 105: 7899-7906.

p. 271 "As Mark Twain..." Twain, Mark. 2004. Letters from the Earth. New York: Harper Perennial. pp. 221-226.

Acknowledgments and Permissions

FIGURE 1 – From Franzen, J.L., P.D. Gingerich, J. Habersetzer, J.H. Hurum, J. von Koenigswald, and B.H. Smith. 2009. Complete Primate Skeleton From the Early Eocene of Messel in Germany: Morphology and Paleobiology. *PLoS One* 4, no. 5 e5723 (May 19) http://www.plosone.org/article /info:doi/10.1371/journal.pone.0005723

FIGURE 2, 3 – Reprinted by permission from Macmillan Publishers Ltd: Seiffert, E., J.M.G. Perry, E.L. Simons, and D.M. Boyer. 2009. Convergent evolution of anthropoid-like adaptations in Eocene adapiform primates. *Nature* 461: 1118-1121.

FIGURE 4 – Reprinted by permission from Macmillan Publishers Ltd: Dalton, Rex. 2009. Fossil primate challenges Ida's place. *Nature* 461: 1040.

FIGURE 5 – From Stensen, Niels. 1667. *Elementorum myologiæ specimen, seu musculi descriptio geometrica : cui accedunt Canis Carchariæ dissectum caput, et dissectus piscis ex Canum genere*. Floerentiae: Stellae.

FIGURE 6 – A portrait of Georges Cuvier painted by 1798 by Mathieu-Ignace van Bree.

FIGURE 7 – From Buckland, William. 1836. Geology and Mineralogy Considered With Reference to Natural Theology. London: William Pickering.

FIGURE 8 – From Gordon, Elisabeth. 1894. The Life and Correspondence of William Buckalnd, D.D., F.R.S. London: John Murray.

FIGURE 9 – From the Images from the History of Medicine (IHM) collection. Order No. B018002. http://ihm.nlm.nih.gov/images/B18002

FIGURE 10 – From Packard, Alpheus. 1901. Lamarck, the Founder of Evolution. New York: Longmans, Green, and Co.

FIGURE 11 – From Gordon, Elisabeth. 1894. The Life and Correspondence of William Buckalnd, D.D., F.R.S. London: John Murray.

FIGURE 12 – A portrait of Charles Darwin, painted by George Richmond in the late 1830's. Reproduced from Wikipedia: http://en.wikipedia.org/wiki/File:Charles_Darwin_by_G._Richmond.jpg

FIGURE 13 – From Lankester, Ray. 1905. Extinct Animals. New York: Henry Holt and Company.

FIGURE 14 – From Scott, W.B. 1913. A History of Land Mammals in the Western Hemisphere. New York: The Macmillan Company.

FIGURE 15 – From Buckland, William. 1836. Geology and Mineralogy Considered With Reference to Natural Theology. London: William Pickering.

FIGURE 16 – From Marchant, James. 1916. Alfred Russel Wallace – Letters and Reminiscences. Vol. 1. New York: Cassell and Company.

FIGURE 17 – From Pearson, Karl. 1914. The Life, Letters, and Labours of Francis Galton. London: Cambridge University Press..

FIGURE 18 – From Darwin, Charles. 1859. On the Origin of Species by Means of Natural Selection. London: John Murray.

FIGUREe 19 – From Waddy, Frederick. 1874. Cartoon Portraits and Biographical Sketches of Men of the Day, Second Edition. London: Tinsley Brothers.

FIGURE 20 – From Thorpe, T.E. 1878. Coal: Its History and Uses. London: Macmillan and Co.

FIGURE 21 – Reprinted by permission from Macmillan Publishers Ltd: Ahlberg, Per Erik, Jennifer Clack, and Henning Blom. 2005. The axial skeleton of the Devonian tetrapod Ichthyostega. *Nature* 437: 137-140.

FIGURE 22 – Reprinted by permission from Macmillan Publishers Ltd: Bosivert, Catherine. 2005. The pelvic fin and girdle of Panderichthys and the origin of tetrapod locomotion. *Nature* 438: 1145-1147

FIGURE 23 – Reprinted by permission from Macmillan Publishers Ltd: Daeschler, Edward, Neil Shubin, and Farish Jenkins Jr. 2006. A Devonian tetrapod-like fish and the evolution of the tetrapod body plan. *Nature* 440: 757-763.

FIGURE 24 – Reprinted by permission from Macmillan Publishers Ltd: Daeschler, Edward, Neil Shubin, and Farish Jenkins Jr. 2006. A Devonian tetrapod-like fish and the evolution of the tetrapod body plan. *Nature* 440: 757-763.

FIGURE 25 – Reprinted by permission from Macmillan Publishers Ltd: Ahlberg, Per Erik, and Jennifer Clack. 2006. A firm step from water to land. *Nature* 440: 747-749.

FIGURE 26 – Reprinted by permission from Macmillan Publishers Ltd: Ahlberg, Per, Jennifer Clack, Ervins Luksevics, Henning Blom, Ivars Zupins. 2008. Ventastega curonica and the origin of tetrapod morphology. *Nature* 453: 1199-1204.

FIGURE 27 – From Hitchcock, Edward. 1858. Ichnology of New England. Boston: William White.

FIGURE 28 – Image from Wikipedia. http://en.wikipedia.org/wiki/File:Huxley7.jpg.

FIGURE 29 – Left image from Goodrich, S.G. 1859. Illustrated Natural History of the Animal Kingdom, Vol. 2. New York: Derby & Jackson. Right Image from Hutchinson, H.R. 1910. Extinct Monsters and Creatures of Other Days. London: Chapman & Hall.

FIGURE 30 – From Huxley, T.H. 1886. American Addresses. London: Macmillan and Co.

FIGURE 31 – From Huxley, T.H. 1886. American Addresses. London: Macmillan and Co.

FIGURE 32 – From Huxley, T.H. 1886. American Addresses. London: Macmillan and Co.

FIGURE 33 – From Heilmann, Gerhard. 1926. The Origin of Birds. New York: D. Appleton and Company.

FIGURE 34 – Beebe, William. 1915. A Tetrapteryx stage in the ancestry of Birds. *Zoológica* 2: 36-52.

FIGURE 35 – From Broom, Robert. 1913. On the South-African pseudosuchian Euparkeria and allied genera. *Proceedings of the Zoological Society of London* 619-633.

FIGURE 36 – Image courtesy Brett Booth.

FIGURE 37 – Restoration courtesy Mark Witton.

FIGURE 38 – Reprinted by permission from Macmillan Publishers Ltd: Hu, Dongyu, Lianhai Hou, Lijun Zhang, and Xing Xu. 2009. A pre-Archaeopteryx troodontid theropod from China with long feathers on the metatarsus. *Nature* 461: 640-463.

FIGURE 39 – Restoration courtesy Mark Witton.

FIGURE 40 — Reprinted by permission from Macmillan Publishers Ltd: Zhang, Fucheng, Zhonghe Zhou, Xing Xu, Xiaolin Wang, and Corwin Sullivan. 2008. A bizarre Jurassic maniraptoran from China with elongate ribbon-like feathers. *Nature* 455: 1105-1108.

FIGURE 41 – Reprinted by permission from Macmillan Publishers Ltd: Xu, X., and M. Norell. 2004. A new troodontid dinosaur from China with avian-like sleeping posture. *Nature* 431: 838-841.

FIGURE 42 – From Sereno, Paul, Ricardo Martinez, Jeffrey Wilson, David Varricchio, Oscar Alcober, and Hans Larsson. 2008. Evidence for avian intrathoracic air sacs in a new predatory dinosaur from Argentina. *PLoS One* 3: e3303.

FIGURE 43 – Reprinted by permission from Macmillan Publishers Ltd: O'Conner, Patrick, and Leon Claessens. 2006. Basic avian pulmonary design and flow-through ventilation in non-avian theropod dinosaurs. *Nature* 436: 253-256.

FIGURE 44 – From Wolff, Ewan, Steven Salisbury, John Horner, and David Varricchio. 2009. Common avian infection plagued the tyrant dinosaurs. *PLoS One* 4: e7288.

FIGURE 45 – Reprinted by permission from Macmillan Publishers Ltd: Hu, Dongyu, Lianhai Hou, Lijun Zhang, and Xing Xu. 2009. A pre-Archaeopteryx troodontid theropod from China with long feathers on the metatarsus. *Nature* 461: 640-463.

FIGURE 46 – From Owen, R. 1861. Palaeontology. Edinburgh: Adam and Charles Black.

FIGURE 47 – From Gould, John. 1863. The Mammals of Australia. London: the author.

FIGURE 48 – Restoration courtesy Matt Celeskey.

FIGURE 49, 50 – Reprinted by permission from Macmillan Publishers Ltd: Luo, Zhe-Xi, Peiji Chen, Gang Li, and Meng Chen. 2007. A new eutriconodont mammal and evolutionary development in early mammals. *Nature* 446: 288-293.

FIGURE 51 – Reprinted by permission from Macmillan Publishers Ltd: Luo, Zhe-Xi. 2007. Transformation and diversification in early mammal evolution. *Nature* 450: 1011-1019.

FIGURE 52 – Reprinted by permission from Macmillan Publishers Ltd: Ji, Qiang, Zhe-Xi Luo, Chong-Xi Yuan, John Wible, Jian-Ping Zhang, and Justin Georgi. 2002. The earliest known eutherian mammal. *Nature* 416: 816-822.

FIGURE 53 – From Fowler, O.S. 1846. The American Phrenological Journal and Miscellany, Vol. 8. New York: Fowlers & Wells.

FIGURE 54 – From Gidley, J.W. 1913. A recently mounted Zeuglodon skeleton in the United States National Museum. *Proceedings of the United States National Museum* 44: 649-654.

FIGURE 55 – Reprinted by permission from Macmillan Publishers Ltd: Gingerich, Philip, S. Mahmood Raza, Muhammad Arif, Mohammad Anwar, and Xiaoyuan Zhou. 1994. New whale from the Eocene of Pakistan and the origin of cetacean swimming. *Nature* 368: 844-847.

FIGURE 56 –From Gingerich, P. D., Munir ul-Haq, Wighart von Koenigswald, William Sanders, B. Holly Smith, and Iyad Zalmout. 2009. New protocetid whales from the Middle Eocene of Pakistan: Birth on land, precocial development, and sexual dimorphism. *PLoS One* 4: e4366.

FIGURE 57 – Reprinted by permission from Macmillan Publishers Ltd: Thewissen, J. G. M, E. M. Williams, L. J. Rose, and S. T. Hussain. 2001. Skeletons of terrestrial cetaceans and that relationship of whales to artiodactyls. *Nature* 413: 277-281.

FIGURE 58 – Reprinted by permission from Macmillan Publishers Ltd: de Muizon, Christian. 2001. Walking with whales. Nature 413: 259-260.

FIGURE 59 – Reprinted by permission from Macmillan Publishers Ltd: Thewissen, J.G.M. 2007. Whales originated from aquatic artiodactyls in the Eocene epoch of India. *Nature* 450: 1190-1194.

FIGURE 60 – From Spaulding, Michelle, Maureen O'Leary, and John Gatesy. 2009. Relationships of Cetacea (Artiodactyla) among mammals: Increased taxon sampling alters interpretations of key fossils and character evolution. *PLoS One* 4: e7062.

FIGURE 61, 62 – Reprinted by permission from Macmillan Publishers Ltd: Gingerich, Philip, S. Mahmood Raza, Muhammad Arif, Mohammad Anwar, and Xiaoyuan Zhou. 1994. New whale from the Eocene of Pakistan and the origin of cetacean swimming. *Nature* 368: 844-847.

FIGURE 63 – From Marino, Lori, Richard Connor, R. Ewan Fordyce, Louis Herman, Patrick Hof, Louis Lefebvre, David Lusseau, Brenda McCowan, Esther Nimchinsky, Adam Pack, Luke Rendell, Joy Reidenberg, Diana Reiss, Mark Uhen, Estel Van der Gucht, Hal Whitehead. 2007. Cetaceans have complex brains for complex cognition. *PLoS Biology* 5: e139.

FIGURE 64 – From Demere, Thomas, Michael McGowen, Annalisa Berta, and John Gatesy. 2008. Morphological and molecular evidence for a stepwise evolutionary transition from teeth to baleen in mysticete whales. Systematic Biology 57: 15-37 by permission of Oxford University Press.

FIGURE 65 – Image from Wikipedia http://en.wikipedia.org/wiki/File:Rembrandt_Peale-Thomas_Jefferson.jpg.

FIGURE 66 – Gould, Charles. 1886. Mythical Monsters. London: W.H. Allen & Co.

FIGURE 67 – From Cuvier, Georges. 1825. *Recherches sur les ossemens fossiles : où l'on rétablit les charactères de plusieurs animaux dont les révolutions du globe ont détruit les espèces*. Paris: G. Dufour et E. d'Ocagne.

FIGURE 68 – From Buckland, William. 1836. Geology and Mineralogy Considered With Reference to Natural Theology. London: William Pickering.

FIGURE 69 – Left image from Andrews, C. W. 1908. A Guide to the Elephants Exhibited in the Department of Geology and Palaeontology in the British Museum. London: Taylor and Francis. Right image from Osborhn, H. F. 1907. Hunting the Ancestral Elephant in the Fayum Desert. *The Century Magazine* 74: 815-835.

FIGURE 70 – From Andrews, C. W. 1908. A Guide to the Elephants Exhibited in the Department of Geology and Palaeontology in the British Museum. London: Taylor and Francis.

FIGURE 71 – From Scott, W. B. 1913. A History of Land Mammals in the Western Hemisphere. New York: Macmillan and Company.

FIGURE 72 – From Gheerbrant, Emmanual. 2009. Paleocene emergence of elephant relatives and the rapid radiation of African ungulates. *Proceedings of the National Academy of Sciences* 106: 10717-10721.

FIGURE 73, 74, 75 – From Shoshani, Jeheskel, Robert C. Walter, Michael Abraha, Seife Berhe, Pascal Tassy, William J. Sanders, Gary H. Marchant, Yosief Libsekal, Tesfalidet Ghirmai, and Dietmar Zinner. A proboscidean from the late Oligocene of Eritrea, a "missing link" between early Elephantiformes and Elephantimorpha, and biogeographic implications. *Proceedings of the National Academy of Sciences* 103: 17296-17301.

FIGURE 76 – From Scott, W. B. 1913. A History of Land Mammals in the Western Hemisphere. New York: The Macmillan Company.

FIGURE 77 – Reprinted by permission from Macmillan Publishers Ltd: Miller, Webb, Daniela Drautz, Aakrosh Ratan, Barbara Pusey, Ji Qi, Arthur Lesk, Lynn Tomsho, Michael Packard, Fangqing Zhao, Andrei Sher, Alexei Tikhonov, Brian Raney, Nick Patterson, Kerstin Lindblad-Toh, Eric Lander, James Knight, Gerard Irzyk, Karin Fredrikson, Timothy Harkins, Sharon Sheridan, Tom Pringle, and Stephan Schuster. 2008. Sequencing the nuclear genome of the extinct woolly mammoth. *Nature* 456: 387-392.

FIGURE 78 – From Gosse, Edmund. 1907. Father and Son. London: William Heinemann.

FIGURE 79 – Library of Congress Prints and Photographs Division. Brady-Handy Photograph Collection. http://hdl.loc.gov/loc.pnp/cwpbh.04124. CALL NUMBER: LC-BH832-175.

FIGURE 80 – Marsh, O. C. 1879. Polydactyle Horses, Recent and Extinct. *American Journal of Science* 17: 499-505.

Figure 81 – From Jordan, David Starr. 1910. Leading American Men of Science. New York: Henry Holt and Company.

FIGURE 82 – Matthew, W. D. 1926. The Evolution of the Horse: A Record and Its Interpretation. *The Quarterly Review of Biology*, 1: 139-185.

FIGURE 83 – From Scott, W. B. 1913. A History of Land Mammals in the Western Hemisphere. New York: The Macmillan Company.

FIGURE 84 – From MacFadden, Bruce. 2005. Fossil horses – Evidence for evolution. *Science* 307: 1728-1730.

FIGURE 85 – From Marsh, O. C. 1892. Recent polydactyle horses. *American Journal of Science* XLIII: 339-355.

FIGURE 86 – From Tyson, Edward. 1699. Orang-Outang, sive Homo Sylvestris. Or, the Anatomy of a Pygmie Compared With That of a Monkey, an Ape, and a Man. London.

FIGURE 87 – From Huxley, T. H. 1863. Evidence as to Man's Place in Nature. New York: D. Appleton and Company.

FIGURE 88 – From Hrdlicka, Ales. 1916. The Most Ancient Skeletal Remains of Man (Second edition). Washington: Government Printing Office.

FIGURE 89 – From Hrdlicka, Ales. 1916. The Most Ancient Skeletal Remains of Man (Second edition). Washington: Government Printing Office.

FIGURE 90 – From Ungar, Peter, Frederick Grine, and Mark Teaford. 2008. Dental microwear and diet of the Plio-Pleistocene hominin Paranthropus boisei. *PLoS One* 3: e2044.

FIGURE 91 – Reprinted by permission from Macmillan Publishers Ltd: Alemseged, Zeresenay, Fred Spoor, William Kimbel, Rene Bobe, Denis Geraads, Denee Reed, and Jonathan Wynn. 2006. A juvenile early hominin skeleton from Dikika, Ethiopia. *Nature* 443: 296-301.

FIGURE 92 – Reprinted by permission from Macmillan Publishers Ltd: Zollikofer, Cristoph, Marcia Ponce de Leon, Daniel Lieberman, Franck Guy, David Pilbeam, Andossa Likius, Hassane Mackaye, Patrick Vignaud, and Michel Brunet. 2005. Virtual cranial reconstruction of Sahelanthropus tchadensis. *Nature* 434: 755-759.

FIGURE 93 – From Klein, Richard. 2009. Darwin and the recent African origin of modern humans. *Proceedings of the National Academy of Sciences* 106: 16007-16009.

Index